EARTH WORK

EARTH WORK

RESOURCE GUIDE TO NATIONWIDE GREEN JOBS

The Student Conservation Association

*Edited by Joan Moody and
Richard Wizansky*

HarperCollins*West*
A Division of HarperCollins*Publishers*

HarperCollins West and the Student Conservation Association, in association with the Rainforest Action Network, will facilitate the planting of two trees for every one tree used in the manufacture of this book.

Since 1957, the Student Conservation Association (SCA) has pursued its mission to foster lifelong stewardship of the environment by offering opportunities for education, leadership, and personal development while providing the highest quality public service in natural resource management, environmental protection, and conservation.

Book Design by Ralph Fowler
Illustrations by Tim A. Clark
Set in ITC Giovanni with ITC Kabel by TBH/Typecast
FIRST EDITION

Library of Congress Cataloging-in-Publication Data
Earth work : resource guide to nationwide green jobs / the Student Conservation Association ;
    edited by Richard Wizansky and Joan Moody. — 1st ed.
        p. cm.
    ISBN 0–06–258543–6 (alk. paper) : $25.00. — ISBN 0–06–258531–2 (alk. paper : pbk.) : $14.00
    1. Environmental sciences—Vocational guidance—United States.   2. Environmental protection—
Vocational guidance—United States.   3. Green movement—Vocational guidance—United States.
4. Pollution control industry—Vocational guidance—United States.   5. Student Conservation
Association (U.S.)   I. Wizansky, Richard.   II. Moody, Joan, 1949-   .   III. Student Conservation
Association.
GE60.E27   1994
363.7′0023′73—dc20   94-3074
                                                                                          CIP

94 95 96 97 98 RRD(H) 10 9 8 7 6 5 4 3 2 1
This edition is printed on acid-free paper that meets the American National Standards Institute Z39.48 Standard.

We call upon Earth, our planet home, with its
beautiful depths and soaring heights, its vitality
and abundance of life, and together we ask that it
Teach us, and show us the way.

—*Chinook Blessing Litany*

The Student Conservation Association dedi-
cates this book to the thousands of volunteers
who have served in SCA programs in parks,
refuges, and other areas around the country; to
"students" of Earth of all ages; and to all who
work to protect the land and the environment.

CONTENTS

FOREWORD

Earth Work: Resource Guide to Nationwide Green Jobs was undertaken to address a question increasingly posed by Americans of all generations: What can I do to help save our environment? Students exploring career possibilities, as well as professionals in search of career changes or meaningful volunteer experiences, are often unacquainted with the opportunities that exist today at both the national and local levels. This guide was produced because we cannot afford to waste the concern or willingness of any American; therefore I hope that it will become a well-thumbed resource in all of our personal libraries.

I want to thank you for contemplating work in conservation. Americans who devote themselves to this important service—the restoration and preservation of our natural resources—deserve the deepest respect and appreciation. It is my hope that, as a result of today's diligent efforts, our children and their children will look back at our lives unashamed and honor our hard work and dedication. You purchase the praise of future generations for all of us.

Theodore Roosevelt IV
SCA Board of Directors

ACKNOWLEDGMENTS

The editors wish to extend their grateful thanks and appreciation to the following friends, colleagues, associates, and significant others who worked to make *Earth Work* possible.

T. Destry Jarvis, Special Assistant to the Director, National Park Service
The Student Conservation Association Board of Directors and Staff
Scott D. Izzo, President, SCA
Shirley E. Christine, Editor, HarperCollins West
Lisa K. Younger, Editor, Earth Work *magazine*
Sherry L. Reed, Circulation Manager, Earth Work *magazine*
Eileen Rodgers
Dr. Todd Mandell
John Alfred Tyson
John Esson and the Environmental Career Center
Kevin Doyle and the Environmental Careers Organization

INTRODUCTION

WELCOME TO EARTH WORK

What is "earth work"? As I sit here, the definition initially seems uncomplicated. Outside my window, an autumn panorama of brilliant reds and yellows rolls over the hills of New Hampshire. The crisp winds and colors of fall elicit associations with our country's incomparable natural resources, which must be maintained and conserved for our enjoyment and use today and for our children's tomorrow. From this vantage point, I define earth work as the traditional conservation careers that young people, many of whom first volunteer with the Student Conservation Association (SCA), enter as park rangers or as scientists and technicians peering into microscopes in fish and wildlife labs. Earth work—in this rendition—sounds as though it should be always happening out-of-doors, close to nature.

But conservation and environmental careers have changed dramatically over the past two decades. Since SCA first began sharing career advice more than thirty-six years ago, passage of environmental laws and the growing public concern about conserving natural resources, and about hazardous waste, water pollution, recycling, global warming, acid rain, tropical deforestation, endangered wildlife, and cultural diversity in the conservation workforce have created a complex variety of new job opportunities. And with these changes, definitions of earth work—both here at SCA and in the world at large—have expanded beyond jobs aimed at conserving land and other resources to include new fields, ranging from environmental cleanup to green marketing. This book is about the richness of meaning that earth work has taken on in our "green" decade of the 1990s.

Today as many as three million people work in environmental industries, nonprofits, and state and federal conservation agencies. More than ten thousand nonprofit environmental groups exist in the United States alone. In the wake of the National Environmental

Education Act of 1990, schools throughout the nation are implementing environmental education programs, and there is an explosion of environmental material in the media.

The earth workers profiled in SCA's *Earth Work* magazine—and now in this resource guide—show us jobs ranging from those much-coveted positions in Yellowstone or Yosemite national parks to the new kinds of environmental management, technical, and communications jobs to be found in government agencies, consulting firms, power plants, treatment facilities, classrooms, and corporations around the nation.

Readers will meet Steve Sarles, who worked his way up from SCA college volunteer and National Park Service seasonal employee to his current position as supervisory ranger for Yellowstone National Park. Jane Nicholich's story will illustrate how she landed her classic dream job with the Fish and Wildlife Service and now cares for the flocks of endangered cranes in her charge at Patuxent Wildlife Research Center.

No matter what your age or background, you will find conservation career goals to pursue in this guide. Consider the example of Carol Collinson Kahl. After ten years in the workforce and a divorce, she launched a new career as a marine scientist by entering graduate school. When *Earth Work* caught up with her, she was working as a Knauss Sea Grant fellow in marine science and policy at the National Marine Fisheries Service, one of the government agencies profiled in this book.

This resource guide will also introduce you to representatives of the new forces in the nonprofit environmental community—resource economists like Jeff Olson, for example. Jeff formerly managed sales for a timber company and now works for The Wilderness Society, where he has been active in countering job-loss claims issued by the timber industry in the Northwest during the debate over national forest protection. Jane Kollmeyer, also profiled here, prefers to use her forestry background working for the "dynamic, changing, and growing" U.S. Forest Service, helping manage a national forest bigger than some states.

The academic backgrounds, job descriptions, and even styles of work covered in this book vary greatly, but each can serve as a model for you as you define and pursue your own earth work career path. For instance, the chapter on nonprofit organizations features Abby Arnold, an environmental mediator employed by RESOLVE, who maintains a firm neutrality between business and environmental groups. By contrast, Bruce Hamilton, whom you will also read about in that chapter, is known for his outspoken advocacy as a top conservation official with the Sierra Club.

If we consider my earlier definition of earth work in the context of these and other professionals profiled in this book, it is clear that a complete description of conservation and environmental careers has to go well beyond the out-of-doors. Today, any definition must encompass all work performed to conserve or restore land, wildlife, and other natural resources or to protect, remediate, or improve the quality of our air, water, or overall environment. It is a definition that boils down to any work performed in behalf of the web of life we call planet Earth or, in short, "earth work." As SCA well knows, much of this work is done on a volunteer basis. But for those of you who wish to make earth work your life's work, this book explores the ways and means to understand and train for your future livelihood in a "green" career.

Whether you are looking for your first job or considering a midlife career change, *Earth Work* magazine's articles and job listings will help you. This book brings you the best of the magazine's career advice—from the philosophical reasons for choosing jobs to the nitty-gritty of how to get your foot in the door and how much money you can expect to make. The articles, surveys, and profiles all have been selected with you—the green career or job seeker or changer—in mind.

Earth Work provides an exclusive summary of graduate school opportunities as well as an entire chapter on salary surveys that includes jobs in both public and private sectors for comparison. You will also benefit from profiles of potential employers—government agencies, nonprofit conservation groups, and environmental management organizations. And you will find a whole chapter on new and "green-hot" environmental careers.

Also included in this guide are job listings from the pages of *Earth Work* magazine (see chapter 2 for a sample clipping). Many people tell me they read these job listings even though they are not currently looking because the listings represent their best source of information on conservation organizations, comparative salaries, and who is doing what.

As you begin to focus in on a specific career from the large variety of natural resources or environmental jobs, I suggest that you seek advice from those currently in the field, not only for descriptions of their jobs but also for insight on what motivates them and what they like about their work. Earth workers can give you constructive tips on what can help you break into conservation—such as service with the Peace Corps, interdisciplinary study, or language training.

Perhaps the most important lesson you will learn in this book is never to give up your goals. Neil Curry, an interpreter at Pinnacle Mountain State Park in Arkansas, says, "People in the field told me that I would never find a job. The jobs are out there; you just have to go look for them. You have to be dedicated enough to volunteer to get in the door."

I hope you will come away from these pages learning that every new career was once just someone's dream. Even as I write this introduction, new "green-hot" careers are in the works. This guide is intended to help you on your way.

Good luck with your job search.

Scott D. Izzo
President, Student Conservation Association

(Please note that the job listings displayed in this book are for jobs that are currently filled. New job listings and career advice are available every month in Earth Work *magazine.)*

Chapter One

THE JOB HUNT IN GREEN

The 1990s will be the decade of green careers. Despite economic and political uncertainties, a convergence of trends makes green jobs hot for the rest of this century and into the next. Public concern and policy actions related to hazardous waste, water pollution, recycling, global warming, acid rain, tropical deforestation, and endangered wildlife have created unprecedented job opportunities.

How big is the green job market, and how big could it get? In the summer of 1992, *Earth Work* magazine put together estimates from environmental industries and government conservation agencies, added some for nonprofits, and gave a conservative total estimate of "more than a million." That total apparently was more conservative than we guessed. Environmental Business International (EBI), based in San Diego, estimates that the U.S. environmental industry employed 1.1 million people in 1992 and will increase 18 percent by 1997. In addition, the U.S. Interior Department employs more than 80,000 people, the Forest Service approximately 32,000, the National Marine Fisheries Service 2,500, and EPA almost 20,000. According to various estimates, about 100,000 people work for state governments in parks, outdoor recreation, forestry, fish and wildlife management, and other land and water conservation jobs, and perhaps 150,000 work for the states in environmental protection and cleanup. The nation's nonprofit environmental organizations probably account for at least another 30,000; environmental educators, communicators, and planners total more than 100,000. Adding those rough figures together yields a total twice our estimate. Other estimates go up to 3 million. In fact, because of the great variety of job descriptions and places of work, no one can tell the exact number of earth workers out there—especially considering that you can make any job greener if you so desire.

More important to you as a job seeker is the question of how big the market will get. In short, everyone agrees that the market will get bigger—fast. Demand is greatest for employees with a technical or scientific background in the private environmental protection industries. Jobs listed each month, however, demonstrate that plenty of opportunities are available for natural resources and liberal arts graduates as well as environmental scientists. The positions are found in federal, state, and local government agencies, private industry, and nonprofit organizations.

Here are ten trends that show how natural resources and environmental careers are changing in the 1990s:

1. GROWTH. A surge of environmental and natural resources jobs is being created in the '90s as more than sixteen federal environmental protection acts and several dozen land protection laws passed since 1970 get implemented during a time of unprecedented public support for the environment.

The U.S. Department of Labor's *Occupational Outlook Handbook* says, "Efforts to preserve the environment are expected to result in growth of jobs in that area in the 1990s." Since that prediction, President Bill Clinton and Vice President Al Gore have been boosting U.S. environmental protection and renewable energy technologies as sources of jobs. They have promised to "shatter the false choice between environmental protection and economic growth by creating a market-based environmental protection strategy that rewards conservation and 'green' business practices while penalizing polluters" (*Putting People First* [New York: Random House, 1992]).

A November 22, 1993, Administration report ordered by President Clinton says the world market for "envirotech" products and services, such as advanced waste disposal, smog mitigation, and water purification, is expected to more than double to $600 billion by the year 2000. The administration says this is "one of the most attractive areas in which to focus the U.S. government's export-promotion efforts." Michael Silverstein, head of Environmental Economics, estimated that as of 1993, there were 65,000 to 70,000 American firms involved in some kind of environmental cleanup, and thousands more "smallish ventures" that make, sell, install, or create environment-saving products or concepts.

Because of the administration's plans to cut jobs in all agencies government wide, growth for the foreseeable future may be mostly in the private sector, with the exception of specific government jobs in environmental protection and cleanup. It is not yet clear how many federal land conservation and environmental protection jobs will be cut or whether the cuts will concentrate on administrative personnel.

Scientists, engineers, lawyers, auditors, fund-raisers, and specialized consultants will continue to have particularly good prospects in both public and private sectors.

Chapter 7 presents some particularly attractive fields of growth. For example, the Environmental Careers Organization (ECO) predicts that there will be 150,000 new jobs in solid waste management by 1995 and 75,000 new jobs in hazardous waste management; ECO also estimates about 15,000 new air quality professionals annually and an equivalent number in water quality. And more recently, ECO predicted that over the next five years all the major environmental industry "tracks" are expected to experience growth. These include solid waste management, hazardous waste management, environmental consulting, air pollution control, and others. "Environmental energy sources" is expected to lead the pack, along with air pollution control; asbestos abatement will bring up the rear. *Overall environmental industry growth between 1994 and 1998 will probably be around 6 percent.*

Chapter 8 covers some jobs that are so new or so promising that *Earth Work* calls them "green-hot" jobs.

2. NEED FOR DIVERSITY. Despite its growth, the environmental movement lacks cultural diversity. Employers are seeking out more women and people of color, who are found in disproportionately low numbers in graduate schools in the sciences as well as in nonprofit advocacy groups. Efforts to provide more diversity are covered in chapter 9.

3. COMPETITION IN NATURAL RESOURCES. In general, more new jobs are available in environmental sciences than in natural resources fields. One of the most coveted jobs in America is that of national park ranger, yet the competition is stiff for this much-glorified but low-wage job. Although the government in general is cutting back, there is a growing demand at some agencies and in the private sector for natural resource managers with science and natural resources management degrees. Many employees are nearing retirement at the Bureau of Land Management, for example. Specializing in environmental impact analysis within a discipline such as forestry or fisheries will greatly increase your chances of success. Women and people of color will find special opportunities in natural resources jobs.

Over the long run, the nation will need many foresters, ecologists, and other natural resource professionals. With the immediate outlook not bright, now is a good time to get a graduate degree, considered necessary at the entry level in competitive natural resources fields.

The competition for natural resources jobs means that young people should seek out internships and opportunities to gain experience, such as working alongside natural resource agency officials as a Student Conservation Association volunteer (see the profile of SCA in chapter 2).

4. THE THREE R'S: REMEDIATION, RECYCLING, AND RESTORATION. Biologists, chemists, geologists, engineers, and others are needed for remediation work such as cleaning up oil spills, for implementing the recycling programs springing up around the nation, and for land restoration. Over the next several years, EPA alone will hire fifteen hundred.

Graduates with training in hazardous waste management will meet only a small percentage of the demand in the 1990s, yet this training is offered at only a few universities.

5. EMPHASIS ON ENVIRONMENTAL EDUCATION. The National Environmental Education Act of 1990 and public interest in the environment have created new opportunities in environmental education and an explosion of environmental material in the media.

6. THE INTERDISCIPLINARY APPROACH. Employers are most interested in job candidates with interdisciplinary work experience or double college majors—natural resources and public policy or economics, for example.

7. THE GLOBAL WORKPLACE. The environmental field increasingly is an international one, as evidenced by the 1992 Earth Summit in Rio and the recent debate over the environmental pros and cons of international trade agreements. Those who combine a science background with study of languages, economics, or international relations have a head start on future careers. And those in traditional careers such as forestry can enhance their prospects by international specialization.

8. THE ENERGY AND ENVIRONMENT LINK. The link between protecting the environment and developing alternative energy technologies has received greater recognition, opening doors to future jobs in technology development and technology transfer to third world countries.

9. JOBS FOR LIBERAL ARTS MAJORS. Despite the need for scientists in all environmental fields, nonprofits and private companies also need environmental fund-raisers and communicators to respond to expanding public interest in the environment. With the environmental education boom, curriculum development will demand liberal arts skills coupled with science.

10. THE NEED FOR PROFESSIONALISM. The book *Inside the Environmental Movement*, by the Conservation Fund, details the need in the nonprofit community for increased professionalism and leadership (see summary in chapter 6).

SIX COMMON QUESTIONS FROM JOB SEEKERS

In this section, the editors have enlisted the help of three experts to give practical advice on job hunting in the conservation and environmental fields and put together their answers to the questions they are most frequently asked.

The three experts have written regular "Green at Work" columns in *Earth Work* magazine, answering mail from readers and passing on the latest tips. They all are knowledgeable about conservation and environmental jobs in general, but *Earth Work* has called on each of them for particular specialties.

John Esson, an environmental career counselor and director of the Environmental Career Center in Hampton, Virginia, has provided us with detailed, user-friendly information on the federal government and currently is preparing a book on the subject. His observations are based on his fourteen years of natural resources and environmental experience with the Bureau of Land Management, Army Corps of Engineers, and other agencies as well as extensive and unique research.

Susan Cohn, a career counselor at New York University's Leonard N. Stern School of Business, specializes in the new business opportunities in the environmental field, such as green marketing and corporate environmental programs. She is the author of *Green at Work.*

Kevin Doyle, as national director of program development for the Environmental Careers Organization (ECO), has been one of the primary authors of ECO's *The Complete Guide to Environmental Careers.* In his role developing programs, including a number of short-term, environmental job opportunities with environmental service firms and other agencies, Kevin brings a special expertise on what is happening in the private, for-profit world of environmental services.

The advice gathered here is augmented by SCA from its own history of involvement with government agencies and knowledge of the nonprofit community.

QUESTION 1: CAN I WORK IN THE ENVIRONMENTAL FIELD AND MAKE A LIVING AT THE SAME TIME?

Environmentalists could be considered the "social workers of the planet." The environmental field is similar to the field of social work in that it frequently requires a period of volunteer work or internship before a salaried job can be obtained. Interning or volunteering can only work in your favor as you seek environmental employment. In other words, get experience working in the field! This is especially true with regard to nonprofit opportunities.

In the nonprofit sector, expect that many salaries are below those of comparable positions in the corporate sector. However, salaries for other jobs—such as fund-raiser—are substantial. Part of choosing a career in this sector is weighing the pluses and minuses of the nonprofit world. The disadvantage may be the low salary, but the advantage may be the opportunity to live your ideals.

Because career options in job sectors other than the traditional nonprofit setting are expanding, more and more are available without having to do internships or volunteering. Career opportunities exist for environmental engineers and for professionals in various job functions that have an environmental component (such as "green" marketing, environmental planning, etc.). There are many routes to explore that provide entrepreneurial opportunities and can command impressive salaries. The choice is an individual one, and one you need to weigh for yourself.

Consider the amount of "ecopreneurial" activity in the environmental field. Millions of dollars are being made from the sale of organic food items, including ice cream, yogurt, candy, and foods, and other products that are "green"; consulting practices are created out of expertise learned on the job. Many jobs have not yet been created. In the future, it will be up to you to see a niche and take advantage of it for your career. As more and more opportunities for employment are in small to medium-size businesses, you have an opportunity to make a difference by bringing your environmental knowledge to a company and instituting programs and policies that work to green the company and the profits.

In addition to ecopreneurial activity, environmental cleanup will be an increasing concern worldwide. Opportunities to clean and green from Mexico to Eastern Europe will provide new career positions for those interested in capitalizing on this growing market sector.

QUESTION 2: WHAT KIND OF COLLEGE DEGREE IS BEST? DO I NEED A MASTER'S DEGREE?

At a recent environmental conference, a room full of environmental professionals was asked about their academic training. The count: thirty-one different college majors. There is no simple academic path to an environmental career. Environmental professionals include chemists, engineers of many stripes, toxicologists, industrial hygienists, geologists, wildlife and fisheries biologists, foresters, soil scientists, and ecologists, of course. Also found among the ranks of "earth workers," however, are archaeologists, anthropologists, filmmakers, journalists, teachers, trainers, photographers, economists, attorneys, managers, accountants, poets, farmers, musicians, laborers, truck drivers, garbage collectors, recycling coordinators, and people from a long list of other trades, disciplines, and professions.

Within such a diverse field, it is virtually impossible to give generic advice about educational preparation. The education one needs to compete for a field research position at the U.S. Geological Survey bears no resemblance at all to the training needed to be a hazardous waste technician, a park ranger, a pollution prevention specialist, or a writer for *Sierra* magazine.

So, how do you decide? First, dream about what you would like to do. There are many books that provide structured exercises for imagining your career dreams, if you need a jump start. Second, identify and meet with people who are doing the work you imagine for yourself. This is not as difficult as it sounds. Once you know the question to ask, you will be surprised at how many people will have suggestions for you. For instance, if you start asking, "Do you know of anyone who makes a living studying coral reefs—and who spends a lot of time scuba diving?" you will uncover a small but fascinating group of environmental scientists, many of whom would be glad to talk to you about their work. Third, ask for advice about academic training. Be prepared for some surprises. You will be pleasantly surprised to find that many people who are happily employed in environmental work may not have the "required" academic qualifications for the jobs they do every day. After you have identified and talked with current environmental professionals who are doing what you might like to do, you will have a wealth of information about possible academic "tracks." Supplement it with what you learn from reading books and magazines, and the answer to this commonly asked question will start to emerge for you.

QUESTION 3: WHAT ARE THE HOTTEST ENVIRONMENTAL PROFESSIONS?

Take a look at the "Help Wanted" ads of any Sunday newspaper. Look under *E* for "Environmental." Whether your Sunday paper is from Denver, Boston, Seattle, Dallas, Tampa, or St. Louis, the list of needed environmental professionals is pretty much the same. Those currently most in demand include the following:

- environmental (and other) engineers
- hydrogeologists/earth scientists
- industrial hygienists
- environmental managers/project managers
- hazardous waste management specialists/technicians
- wetland ecologists
- geographic information systems (GIS) specialists
- environmental chemists
- environmental scientists
- toxicologists/risk assessment specialists

See a pattern here? The environmental professions most in demand are overwhelmingly technical and scientific in nature, with engineers, chemists, and earth scientists leading the way. This does not mean that there are not thousands of nontechnical "environmental" employees being hired. There are. The greatest demand, however, is clearly for the techies.

It is interesting to note that most environmental employers want a well-rounded technical employee, though. The ideal environmental manager, for instance, would be technically competent, a superb communicator, skilled in the management of people (especially multidisciplinary teams), able to schedule and coordinate complex projects, and comfortable around budgets and financial considerations.

For predictions on "green-hot" careers in the future, see chapter 8.

QUESTION 4: I WANT TO WORK IN THE ENVIRONMENTAL FIELD, BUT HOW DO I GO ABOUT IT?

In your job search, you must understand four things; namely, your values, your skills, how to research, and the importance of networking.

VALUES. Before you can market yourself and get a job in the environmental field you must understand yourself. You will need to ask yourself what it is about environmental issues that motivates you to make a contribution and to seek employment with this focus. You will need to ask yourself where you see your values meeting the values of your potential employer. Additionally, you will need to single out the areas to which you are most committed. For example, do you want to work on clean air and water issues, on protecting the rain forest, or on issues of recycling, waste management, or product manufacturing? Asking these questions will assist you in narrowing your job search to specific organizations.

SKILLS. You will need to assess your skills and talents and how they can be utilized in your job search. Also, you need to know how you can contribute to an organization and why they should hire you above and beyond your environmental interest alone. You need to demonstrate a skill base that makes you valuable. For instance, having skills in editing and publishing could meet the needs of a nonprofit environmental research and publishing group. Assessing your skills, along with understanding the goals of an organization, will help you sell yourself to the organization. It is crucial for you to be able to demonstrate what your skills are and why the organization needs you. Being strategic in understanding what you want, what the organization wants, and what you can do together can provide a competitive advantage as you seek employment.

RESEARCH. In order to narrow your job search, you will need to know who is taking the lead in areas of interest to you. Researching companies helps narrow your job search and prepares you to network. Research prepares you to understand how your needs can meet the needs of the companies you have targeted. A part of your research may involve meeting with professionals for an informational interview. In such an interview the focus is not on getting a job but on better understanding the skills and requirements of a specific job function. Informational interviews are an opportunity to meet with professionals in the environmental field and further narrow your search.

NETWORKING. Chances are, if you are not networking, then you are not working: 85 percent of all jobs are found through networking, not through executive search or advertisements. Networking is the key to finding employment. The first step is understanding your network; it consists of your family, friends, friends of friends, and colleagues. Clipping newspaper articles and obtaining names of people involved in your career of interest is also part of it. The way to find employment is to investigate, research, and talk to people to find out who's who in the environment. Read and stay current. Networking is crucial to the career search. When you have assessed your skills and values and done your research and networking, you will be in a position to seek work that is "green."

Green, Green Everywhere

The major areas of increased employment opportunities lie in the environmental services: solid waste management, asbestos removal, water treatment, and pollution control. However, due to certain pressures, what is an "environmental" career is being redefined. Some 85–90 percent of the Fortune 500 and 100 companies contacted by the career office at the New York University Stern School of Business have a "green" or environmental affairs component. Consumers are demanding environmental responsibility and choosing products based on company performance and commitment to environmental issues. Today one can have an environmental specialization within any number of chosen professions. Marketing is one example; many corporations are now concerned with how to effectively market a "green" image. Being environmentally literate will provide a competitive advantage as you seek employment in any sector of the economy.

QUESTION 5: WHAT KINDS OF JOBS ARE OUT THERE FOR ME, IF I DON'T HAVE A SCIENTIFIC OR LAW DEGREE?

Many positions in all industries are now requiring environmental knowledge. Environmental literacy is becoming an important asset in your skill base. What is important is understanding your skills and the areas of work that interest you. For instance, if your love is marketing, you may identify work in nonprofit organizations in which you market the organization's attributes to potential members. You could also take your interest in marketing to a company that makes environmental products or provides environmental services to its clients. Or you may want to use your marketing skills in a corporation, where you can add an environmental overlay to your decisions with regard to packaging, marketing promotions, and campaigns, including possible affiliation with nonprofit organizations. Any skill set can be applied to environmental employment if you understand that environmental imperatives are shaping business practices from an efficiency standpoint as well as from a consumer interest standpoint.

Take any job and green it. Create a niche for yourself as one who has skills in a given area and can also provide ecological understanding to a specific field. Money can be made in "green" retailing, fashion design including environmental components, in architecture; the design professional can choose products to reduce energy use and toxic materials. Any profession, from accounting to finance, to consulting, to entrepreneurship can be lucrative and sustainable for you and your family as well as for our natural resource base.

QUESTION 6: WHAT CAREER OPPORTUNITIES ARE AVAILABLE THROUGH ENVIRONMENTAL RECRUITERS?

One of the major reasons for the surge in jobs and the need for recruiters is the increased enforcement of major environmental laws, including the Superfund Amendments and Reauthorization Act (SARA), the Resource Conservation and Recovery Act (RCRA), and the Clean Air Act (CAA) amendments of 1990. SARA and RCRA have increased the demand for environmental professionals steadily over the past six years. According to a recent article in the *Environmental Reporter*, the Environmental Protection Agency (EPA) estimates that the CAA amendments will add up to sixty thousand new environmental jobs this decade.

Concurrently, the U.S. population is aging. As the young adult population declines, fewer college graduates will enter science and engineering professions. The demand for environmental jobs will continue to increase while the pool of professionals decreases. Employers are having a much more difficult time filling some of their vacancies, so they hire recruiters to help fill the gap between supply and demand. Environmental consulting firms, architectural and engineering firms, and private industrial companies are just some of the businesses that might use recruiters to search for the right individual—someone with both the technical skills and the right corporate "chemistry" for the job in question.

Many of these private for-profit employers use recruiters to supplement their own recruitment departments, or to screen applicants. Some may hire recruiters to discreetly woo away the best personnel from their competitors. Recruiters are private consultants who work for the client companies under contract, with clients paying the entire search fee (usually based on the prospective salary for the vacant position). The service is always free to job seekers. Recruiters need individuals with professional degrees and three or more years of environmental experience. Environmental specialists and engineers with ten or more years' experience are in greatest demand.

Getting Ahead with a Recruiter

Job recruiters can often give you an edge on finding the right position, or finding a better one. Often, they can supply you with unadvertised positions, or positions offering higher salaries (most jobs handled by recruiters range from $35,000 to $100,000 and up). Recruiters can also keep your job search discreet. With a recruiter, you can review a career opportunity, ask questions, and still remain anonymous until you are convinced it is the right job for you. Only then does the recruiter introduce you to the prospective employer.

Once the job search is under way, recruiters keep working for you, especially if you are a top candidate for an open position. The recruiter will help you learn more about the vacancy and give you details on the company, job atmosphere, benefits package, advancement opportunities, surrounding community, and whatever else you need to know to help you make the right career decision.

How to Reach Recruiters

There are more than one hundred recruiting firms that can help your career soar if you have the right background and are flexible about where you will live. Recruiting agencies sometimes put out calls for résumés in technical journals such as *Pollution Engineering* and *HazMat World.*

Job fairs are good places to find recruiters. And remember that recruiters may contact you directly if you have published articles or presented papers at conferences relevant to specific positions, or if you are listed in environmental personnel directories or are a member of a trade association or society.

Before you send out your résumé, contact several recruiting firms and ask specific questions about opportunities in your field. If possible, link up with the recruiter who seems most knowledgeable about your type of job, and who represents the most appropriate clients, in terms of both job opportunities and your geographical preferences. In this way, you can increase your chances of getting the environmental job that is just right for you.

Chapter Two

HOW TO BUILD A GREEN NETWORK

According to recent sources, 85 percent of the job market is "hidden": All the positions advertised through job search firms, newspaper and magazine advertisements, and college placement offices represent only 15 percent of all the jobs out there. It is networking that provides 85 percent of the people with their jobs. The "net" is the matrix of people you know who can assist you with employer contacts. Each is a "source." This net or matrix can also be a tremendous reservoir of information and encouragement to begin your job search.

The best way to start building your network is to identify a list of sources. Be inventive—everyone knows someone who can assist in the job search. Below are a few suggestions.

- ⤳ Friends and work colleagues
- ⤳ Family members and their friends
- ⤳ Organizations you belong to, such as churches, synagogues, special interest groups, and political and environmental organizations
- ⤳ College friends, faculty, and student groups and alumni events
- ⤳ Newspapers, books, magazines, and directories
- ⤳ Lectures, conferences, and events that have an environmental focus

Here are some points to remember as you build your contact base:

- ⤳ When networking, you are requesting information and ideas, *not* asking for a job.

- Most people are flattered to talk with someone interested in their career and will share their knowledge with you.

- Whether you choose to call or to write to initiate networking depends on how you were referred to the person. When in doubt, write a letter and follow up with a phone call to arrange a convenient time to meet.

- Create a tracking system for organizing your contacts. Make sure you have accurate names, phone numbers, and addresses. Keep track of information and additional contact people given to you by those with whom you have met. This tracking system will provide a synopsis of your networking process and can highlight what avenues have been successful.

- When meeting a contact, make the best use of his or her time as well as your own. Remember, this is not a job interview. You are there to gather information. Keep in mind that you are the one who initiated contact for information purposes. It is your job to keep the flow of information coming. Here are some of the questions you may want to ask:

  1. *Could you describe your career path to date and how you entered this field?*

  2. *What are the entry level positions in this field?*

  3. *What is a typical workday like in this position?*

  4. *What do you like or dislike about your job?*

  5. *How would you describe the environment in which you work?*

  6. *What should I stress about my background when I interview in this field?*

  7. *What do you see as your career options after this job?*

- When networking with someone, make sure to ask whether or not there is anyone else in the company or field with whom you could speak. Follow up on every lead!

- Send thank-you notes and follow-up letters.

NETWORKING RESOURCES

An essential component of networking is building your people contacts and resource base. The resources listed in this section are just a sampling of those that have been useful to people networking in the environmental field.

BOOKS AND POSITION PAPERS. Island Press has comprehensive catalogs of environmental books. Topics range from poetry to thorough analysis of major environmental challenges. Contact Island Press to request a catalog and order books. Organizations such as the Natural Resources Defense Council, the Sierra Club, and the World Resources Institute also publish many books and position papers. They are excellent resources for researching varied topics of environmental interest. You can contact them at the following addresses:

Island Press, Box 7, Dept. 4C2, Covelo, CA 95428

Natural Resources Defense Council, 40 West 20th Street, New York, NY 10011

World Resources Institute, 1709 New York Avenue NW, Washington, DC 20006

Sierra Club Books, 730 Polk Street, San Francisco, CA 94109

Worldwatch Institute, 1776 Massachusetts Avenue NW, Washington, DC 20036

PERIODICALS. There are numerous magazines covering environmental topics. *Earth Work* magazine is the best source of career advice, networking information, and job listings, but you will also find it useful to subscribe to other publications that cover specific environmental issues of interest to you. Almost all conservation and environmental membership organizations publish a magazine or newsletter. Magazines of possible interest are

Earth Work, P.O. Box 550, Charlestown, NH 03603, (603) 543-1700

EPA Journal, Superintendent of Documents, Government Printing Office, Washington, DC 20016

In Business: The Magazine for Environmental Entrepreneurship, The HG Press, Emmaus, PA 18048

In addition, a number of newsstand publications provide good background information on general trends:

E Magazine, 28 Knight Street, Norwalk, CT 06851, (203) 854-5559

Earth Journal, 2305 Canyon Boulevard, Boulder, CO 80302, (303) 442-1969

Garbage, 12 Main Street, Gloucester, MA 01930, (508) 283-3200

ENVIRONMENTAL ORGANIZATIONS. A number of organizations are noted for their involvement in various areas. You can contact them for information and decide for yourself which ones you want to join. Also check your local grassroots and regional organizations to request membership information. The best source of information on these groups is *The Conservation Directory* of the National Wildlife Federation. It also includes a number of national, state, and local organizations and government agencies.

GOVERNMENT AGENCIES. Check your local government agencies, such as state departments of parks and recreation, sanitation, and environmental conservation, and health, and the Environmental Protection Agency. These agencies often have publications on health, recycling, outdoor activities, and environmental topics that may interest you.

EDUCATIONAL INSTITUTIONS. Call your local community colleges and universities to request information on any courses that are focused on environmental issues. Environmental courses may be found in many different areas: the sciences, education, business, law, health, and public policy. Courses will broaden your knowledge and provide a forum to discuss environmental issues and meet people with similar interests. Both professors and fellow students become a networking base as well.

NEWSPAPERS AND ELECTRONIC MEDIA. Read newspapers and watch for special television programming on environmental issues. Environmental columns appear regularly in the *New York Times* and the *Wall Street Journal.* Keep abreast of local news in your area as well as national and international issues. Clip articles and keep a file. You may want to contact someone mentioned in an article; making reference to the article when writing someone gives a context for your initial "cold" call or letter.

These suggestions will help you begin to create the support and the knowledge base necessary to find work. As in any job search, the "green" network you create is the key to a successful job search.

ORGANIZATIONS THAT CAN PREPARE YOU

One of the smartest ways to prepare for a green job or career is to start your education early with work experience—paid or volunteer. Each year, thousands of opportunities exist for high school and college students as well as others to participate in programs that accomplish significant conservation and environmental protection work while providing opportunities for self-discovery and environmental education. Perhaps most important, these experiences furnish the insight and skill building that can help you decide which career path you will choose—and to get well ahead in blazing your conservation or environmental career trail.

Four major organizations are leaders in preparing people of all ages for green careers. The first three of the following profiles focus on internships and work experiences with the Student Conservation Association (SCA), the Environmental Careers Organization (ECO), and the Peace Corps. The next profile—of the Environmental Career Center—presents an organization that provides expert advice on conservation careers. Each of these organizations can be used to head you in the direction of your green career dreams—through real work, hands-on experience, and reliable information.

THE STUDENT CONSERVATION ASSOCIATION

When 21-year-old Elizabeth Cushman Titus walked into the office of National Park Service Director Conrad Wirth in the fall of 1954, she had no idea that his response to her proposal would be not only positive but enthusiastic. The proposal was based on the undergraduate thesis Liz was writing at Vassar College, in which she explored the idea of fielding student volunteers for maintaining and protecting America's national parks. At the time, "people were loving the parks to death," and park rangers were finding it difficult to stem the tide of tourism and neglect. Using the model of the Civilian Conservation Corps of the 1930s, the thesis proposed that a national student conservation corps could be formed that would provide volunteer labor for the conservation of national parks while the students took part in conservation education and personal discovery experiences.

The proposal to Director Wirth marked the birth of the Student Conservation Association (SCA) and a relationship that, for the past thirty-seven years, has grown consistently more productive for SCA and the National Park Service and now includes a congressional appropriation that strengthens the association's capacity to provide volunteers for natural resource management and interpretive work in the parks. Elizabeth Titus's contribution to the Park Service was acknowledged in 1989 when she was made an honorary park ranger.

Beginning with fifty student volunteers at Grand Teton and Olympic National Parks in 1954, SCA has become the nation's largest provider of long-term conservation volunteers. Today, SCA fields two thousand participants annually who take part in protection of endangered species, air and water quality monitoring, interpretation, ecological restoration, and related conservation projects. In addition, SCA volunteers build and maintain trails, and—through the Henry S. Francis, Jr., Wilderness Work Skills Program—the association serves as an important center for the collection of information and training in traditional wilderness work skills. The SCA High School Program fields crews of young people on backcountry conservation projects for five- to six-week stretches, and the college-age Resource Assistant Program places volunteers as professional assistants in twelve- to sixteen-week "internships" with resource management personnel. SCA's mission is to accomplish significant conservation work and to present volunteers with opportunities for conservation education and self-discovery.

As SCA programs expand, the association has a firm commitment to stay on the cutting edge of improvements in conservation techniques and environmental education. For instance, ecological restoration has moved to the forefront of today's natural resource management technology, and SCA has been in the vanguard of hands-on work in this arena. In 1989, SCA mobilized a volunteer work force in Yellowstone National Park to restore fire damage caused by the fires and fire fighting efforts in the park as a result of the wildfires of 1988. Subsequently, the Service and SCA joined in a three-year collaboration—The Greater Yellowstone Recovery Corps—which included more than six hundred volunteers who cleared and rebuilt miles of trails and bridges and reestablished ecologically damaged areas.

SCA's commitment to diversity in the conservation workforce has led the association to launch a major national initiative to train young women and people of color for conservation careers. The Conservation Career Development Program (CCDP) guides these young people from the junior year in high school, through college, and into a conservation job. The goals of the CCDP are to expose participants to a variety of intensive conservation opportunities over several years in order to increase their experience and job readiness and to increase representation by women and people of color in the conservation workforce.

And because environmental education and career guidance are at the heart of the "green" decade of the 1990s, SCA is pursuing new and dynamic projects for participants and the conservation community at large. In 1991, for instance, SCA founded *Earth Work*, a monthly magazine that focuses on work in conservation and those who do it. In addition to articles and features, *Earth Work* each month includes a directory of job openings in the conservation and environmental fields (see partial sample listings on next page). The association will continue to explore new ways to provide opportunities for both personal development and the highest quality public service by developing an ongoing assessment of the impact of hands-on conservation experiences on participants. In addition, SCA will continue to use its experience and expertise to publish information, such as that contained here, that can guide people of all ages as they seek out conservation careers and volunteer opportunities. And programs like the CCDP will serve as models for developing tracking systems to monitor conservation career training.

These new initiatives and the association's traditional programs serve the dual mission of SCA since its beginning: to cultivate people while conserving precious natural resources.

The Student Conservation Association, P.O. Box 550, Charlestown, NH 03603, (603) 826-4301

THE ENVIRONMENTAL CAREERS ORGANIZATION

The year was 1972. Richard Nixon was president; the Environmental Protection Agency was just two years old. The words *Love Canal, Times Beach, Bhopal,* the *Exxon Valdez, curbside recycling,* or the *snail darter* had no special meaning. And the Massachusetts Audubon Society gave a $5,000 grant to a recent college graduate for an "environmental intern program." Eight students were placed.

Fast-forward twenty years. Environmental careers now rank as some of the fastest growing professions in the United States, employing nearly two million Americans on expenditures of up to $200 billion a year. Terms like *global warming, acid rain, hazardous waste, "not in my backyard,"* and *spotted owls* are part of the national vernacular. And that $5,000 Massachusetts program has grown into the Environmental Careers Organization (ECO), placing hundreds of new environmental professionals all over the country each year.

ADMINISTRATION

ADMINISTRATIVE ASSISTANT
Conservation Environment and Historic Preservation, Inc., DC

Description: Provide back-up support to other staff; coordinate project activities with senior program directors; maintain division files; and photocopy..

Salary: $20,000 or above, commensurate with experience.

Qualifications: Must have a minimum of two years experience, excellent oral and written communication skills, professional telephone skills, research and organizational skills, and the ability to type 45 words per minute.

To Apply: Send resume and cover letter to S. Villa, Box 18364, Washington, DC 10036. Deadline is February 15.

ADMINISTRATIVE ASSISTANT
Conservation Environment and Historic Preservation, Inc., DC

Description: Assist bookkeeper in preparation of monthly financial analysis reports, accounts receivable and payable, and in maintenance of check books and payroll information; handle logistics for meetings and press conference/briefings; order office supplies; maintain and distribute company manual; and handle phone calls.

Salary: $20,000 or above, commensurate with experience.

Qualifications: Must have a minimum of three years experience, some accounting knowledge, computer skills with knowledge of WordPerfect 5.1 and accounting software; professional telephone skills and strong organizational skills; and ability to type 45 words per minute.

To Apply: See previous CEHP listing (administrative assistant).

ADMINISTRATIVE DIRECTOR
Wild Canid Survival & Research Center, MO

Description: Coordinate the daily operation of the WCSRC involving fund-raising, financial management, and relations with the media, the public, members, the board of directors, volunteers, etc. Must supervise a small staff, assist in production of newsletter, and prepare grant proposals.

Salary: Negotiable.

Qualifications: Must have master's degree or equivalent experience and strong abilities in communications, personnel management, and community relations.

To Apply: Send resume and the names of three references to Attn: Search Committee, WCSRC, Box 760, Eureka, MO 63025. Deadline is March 31.

JOB SCAN

ASSISTANT DIRECTOR
Linsly Outdoor Center, PA

Description: Supervise the overall operation of facilities and programs. Responsible for recruiting and training staff; overseeing risk management practices; maintaining low and high ropes courses; developing curriculum; and planning and implementing Adventure Based Counseling programs.

Salary: Depends on experience.

Qualifications: Should have experience in the following areas: administration (at least three years); outdoor activities such as ropes course, initiatives, rock climbing, caving, extended wilderness trips, and white water canoeing; and developing environmental education programs. Excellent and proven skills in writing and verbal communication,

and marketing and grant writing required. WSI, Advanced First Aid, and CPR Certification recommended.

To Apply: Contact Gregg Somerhalder, Director, Linsly Outdoor Center, 2425 Rte. 168, Georgetown, PA 15043; (412) 899-2100.

ASSOCIATE DIRECTOR OF DEVELOPMENT
The Nature Conservancy, CA

Description: Identify, cultivate, and solicit major gifts and grants from individuals, foundations, and corporations. Supervise several members of the staff; assist in planning and executing capital campaigns; maintain and enhance all development infrastructure, including donor research, maintenance of records, the donor information system, and correspondence; and help develop annual and long-range development plans and strategies.

Salary: Commensurate with experience and qualifications. Includes benefit package.

Qualifications: Must have at least three years experience in management and development with a nonprofit organization with multi-faceted development programs; a proven track record in major donor solicitations; and a strong demonstrated commitment to conservation. Strong interpersonal, leadership, and communication skills; ability to relate comfortably to business and community leaders and to serve as a strong and effective advocate for TNC; and a willingness travel required.

To Apply: Contact Patti Brady, The Nature Conservancy, California Regional Office, 785 Market St., San Francisco, CA 94103.

CONSTRUCTION MANAGEMENT FIELD ENGINEER
Geraghty & Miller, Inc., AZ

Description: Write for details.

Salary: Excellent starting salary and comprehensive benefits.

Qualifications: Must have a BS in construction management or civil engineering and at least two to five years experience in construction management/contractor oversights. Must show good judgement and responsibility and thrive on activity. Good communication and documentation skills a must.

To Apply: Send resume in confidence to Dr. Richard Tinlin, Corporate Technical Recruiter, Geraghty & Miller, Inc., Box 1069, Camp Verde, AZ 86322.

DEPARTMENT HEAD
4-H Farley Outdoor Education Center, MA

Description: Write for details.

Salary: $185 - $200/ week plus room and board, and medical coverage.

"We've gone through a lot of changes and several names," says founder and president John R. Cook, Jr. "We've been EIP, the Center for Environmental Intern Programs, and the CEIP Fund. Today's name, Environmental Careers Organization, says it all, though. It really tells the whole story."

"The whole story" is one of remarkable growth and change. ECO offers a variety of programs and services for environmental job seekers, employers, academics, and the general public. For twenty years, the only constant at ECO has been a commitment to the environment and the people who work to understand and preserve it.

Just what does ECO do? According to its mission statement, the group "is dedicated to protecting and enhancing the environment through the development of professionals, the promotion of careers, and the inspiration of individual action." The organization functions as a national nonprofit organization with four related services and regional offices in Seattle, San Francisco, Cleveland, Tampa, and Boston.

Environmental Placement Services

The best known of ECO's programs is Environmental Placement Services (EPS). More than 5,500 "associates" have gotten their professional start through EPS since that first small group was placed in 1972. This year, more than 550 people will carry out challenging, professional-level projects at private and public environmental employers, called "sponsors," throughout the nation.

"We pride ourselves on offering positions with the whole environmental community," says ECO vice president Mike Rodrigues. "In 1994, more than a hundred different organizations will sponsor our associates, including Fortune 500 corporations, government agencies at all levels, grassroots and mainstream advocacy groups, and major consulting firms. We'll have positions for engineers, scientists, technicians, journalists, policy people, and liberal arts interns."

Environmental Placement Services is not quite like any other "intern program." First, people need not be students to apply, and summer positions are only a small part of the total. It is not unusual for positions to entail nine months or more of full-time work. Also, ECO associates are all paid, sometimes quite well. "Stipends are rising," says Rodrigues. "This year, the range will be from $300 to $700 per week or more; $450 a week is pretty common."

Where does the money come from? Sponsoring organizations pay most of the direct costs, including all of the stipends and payroll taxes. Sponsors also pay a program fee for each associate to help defray the cost of national recruiting, a rigorous screening and matching process, and after-placement services for both associates and sponsors. Remaining program costs are covered through gifts and grants from foundations, corporations, and individuals.

ECO's 1994 sponsors read like a who's who of the environmental community: federal agencies like the EPA, Bureau of Land Management, Department of Energy, and U.S.

Geological Survey; corporations such as IBM, Chrysler, Boeing, BP America, and Polaroid; consulting firms, including Arthur D. Little, R. W. Beck, and CH2M Hill; city agencies in Seattle, San Francisco, and Tampa; and nonprofit groups such as the Trust for Public Land, Natural Resources Defense Council, and others.

Environmental Career Services

ECO's associate placements remain the heart of the ECO, but a variety of newer ventures are what turned the humbly started "environmental intern program" into the many-faceted Environmental Careers Organization. "After so many years in the environmental career field," says Kevin Doyle, ECO director of program development, "we wanted to find ways to serve more people and get the word out about environmental careers. A few years ago we put on the first National Environmental Careers Conference with the Student Conservation Association. Since then, things have really grown."

The result was the establishment of another branch of ECO, Environmental Career Services (ECS). The aforementioned conference, now in its tenth year, remains the key event on the ECS calendar. The conference draws hundreds of students and job seekers from throughout the nation. In addition, ECO regional offices sponsor panel discussions and workshops on career opportunities addressing problems of air quality, water quality, solid waste, hazardous waste, and more.

Over the years, the Environmental Careers Organization has gathered enough information and knowledge to fill a book—so they wrote one. *The New Complete Guide to Environmental Careers* is the flagship product of ECO's third service area: Environmental Career Products. Since its publication, the guide has become the key resource for people seeking help in looking for environmental jobs and is now used as the text in several college classes. ECO staff will be producing more books, reports, and directories for your environmental bookshelf, including an updated version of ECO's 1990 publication, *Becoming an Environmental Professional*.

The newest ECO venture is designed to investigate and report on issues of concern to the environmental community and to help environmental employers with human resource needs. Dubbed "Environmental Research and Consulting," this unit has completed a study of how environmental nonprofit organizations are dealing at all levels with the urgent need for greater racial and ethnic diversity in the environmental movement. ECO researchers also sought ways to involve retired engineers and scientists as consultants to grassroots groups to improve their efforts to reduce use of toxics, resulting in a new ECO program.

One of ECO's greatest strengths is its ability to spot emerging trends and take a leadership role in the environmental community. The group's Diversity Initiative is a case in point. "People of color are dramatically underrepresented in the environmental movement," says DI director Mariella Tan Puerto. ECO has responded with a variety of services, including an environmental summer associate program for minorities that will place over one hundred students of color this year, an environmental grants system to identify and fund minority internships at nonprofit groups, a national minority environmental careers conference, and other projects. "We're trying to correct some of the myths about people of color and the environment," Puerto explains. "When people say that good candidates of color for environmental jobs aren't out there or that people of color are not interested in environmental quality, we can help them see that that's just not true."

As ECO enters its third decade, president and founder John Cook reflects that there is more to be done than ever before and there are more people interested in getting it done. Developing partnerships with others in the field has become more urgent. According to Cook, ECO has recently begun to talk to the leaders of other organizations, such as the Student Conservation Association, who are concerned about the future of environmental careers. "I think you'll be seeing a greater emphasis on our part to work with others and forge common agendas with business, advocates, and government. We see this as a big part of our future," Cook says.

SCA and ECO began discussions about greater collaboration, because both organizations realized that they have similar goals and objectives but are serving largely different constituencies and could maximize the impact of their similar programs by joint initiatives. Although operationally different, the similarities of the career services that both programs offer have drawn them together into a new level of collaboration, adding strength and effectiveness to the nonprofit environmental movement.

Environmental Careers Organization, 286 Congress Street, Boston, MA 02210, (617) 426-4375

PEACE CORPS PASSPORT

The goal of the United States Peace Corps has not changed since its establishment by Congress in 1961: to help developing countries meet their needs for skilled people and to help promote mutual understanding between the people of the United States and the people of developing countries.

What has changed is an increased emphasis on conservation programs. "The environmental field in the Peace Corps . . . is currently receiving a great deal of emphasis," explains Cathy Moser, an environmental specialist with the Peace Corps' Office of Training and Program Support (OTAPS). "There are fifty-five programs with over seven hundred volunteers worldwide." She says opportunities are plentiful and diverse, with "volunteers working as national park managers in Paraguay, game rangers in national parks in Botswana, and soil conservationists on all three continents where we have people." Programs include park and wildlife management, forestry, freshwater fisheries maintenance, agriculture extension, and natural resources development and conservation.

Peace Corps service provides volunteers with experience they are not likely to gain in the United States. Upon returning home after their two-year stint (sometimes longer), they frequently discover this experience is very marketable, especially to the federal government. "There are probably volunteers in every single government agency you can name," says Drew Burnett of OTAPS. "This especially applies to people working in the area of natural resources. Probably most of them are working for the federal government or for nonprofits."

Many National Park Service employees boast a résumé that includes time as a Peace Corps volunteer in parks and wildlife management. Current programs in this field include park planning, park management, resource management and protection, park interpretation, fauna survey, wildlife management planning, and game ranching.

Bill Fitzpatrick entered the Peace Corps in 1987, after working at Olympic National Park in Washington, and became a park planner at Outamba-Kilimi National Park in Sierra Leone on the west coast of Africa. His projects included building twenty miles of new trails into elephant habitat as well as a backcountry ranger station and four bridges for motor vehicles. Fitzpatrick also helped establish and acquire training for anti-poaching patrols.

Fitzpatrick and others like him often find that techniques for managing parks and wildlife in other countries differ substantially from those in the United States. In many African countries, for example, the creation of national parks and protected areas often isolates rural populations from wildlife populations upon which they have traditionally depended for food and other products. Many of these people continue to hunt illegally in the parks. This was true in the Outamba-Kilimi region. "Their concept of wildlife is totally different from ours," Fitzpatrick explains. "If it moves and it's not a pet, they will shoot and eat it."

To resolve such problems, volunteers help devise ways to make wildlife contribute to the local economy so people have an incentive to protect the environment. "We worked with local people to promote the success of the park," Fitzpatrick recalls, "to get to the point where it would be accepted. This involved explaining the advantages of the park. Every six months or so we'd have an open house for the local people. We'd have a big dinner; we'd have music—really a big party that we used to get information to people. We also stressed that projects such as the bridges were intended to benefit both the park and the local population."

Jay Udelhoven, Peace Corps Volunteer
Earth Worker

Student Conservation Association (SCA) alumnus Jaye Udelhoven seemed to be "one of the Peace Corps' busier volunteers," says co-worker Mark Ziminske. Formerly an SCA resource assistant with the Hawaiian Forest Birds Research Project, Jaye joined the Peace Corps directly after graduating from college with a 4.0 GPA and has been very successful in his environmental career. "Jaye's always been interested in the out-of-doors and nature," says Jacque Udelhoven, Jaye's mother. Jaye's versatility has enabled him to travel to many exotic parts of the world to do an array of jobs from the truly mundane work of procuring tools and materials to the exraordinary task of constructing an interpretation center.

Called hardworking and capable of assuming responsibility wherever he is needed by his superiors, Jaye has been chosen to represent the Peace Corps at numerous conferences, including the Wildlife Conservation International Conference on Eco-Tourism in Kenya and the Afro-Montange Conference in Bujumbara.

continued

Global Forestry

Another major conservation emphasis in the Peace Corps is forestry. Some six hundred volunteers currently serve in community forestry projects in fifty-five countries, working with village groups or individual farmers to develop strategies for long-term reforestation efforts. Projects include nursery development and management, agroforestry, reforestation, village woodlots, forest resource management, and introduction of wood-conserving technologies.

John Schubert had been a Student Conservation Association (SCA) resource assistant in the North Cascades National Park in Washington in 1976, assisting backcountry rangers with an experimental revegetation program. From 1983 to 1985, he was a Peace Corps volunteer with his late wife, Pam Matthews, in the Fiji Islands. "It was a stereotype of how people envision the Peace Corps: living in a small village, working with very traditional people," recalls Schubert.

Together, Schubert and Matthews worked in a Rural Community Development Program, essentially a planning program. The intent was to gather information and work with the villagers to assess their needs, possibilities, and resources with an eye toward developing a village plan. Schubert explains, "It was intended as an opportunity for the villagers to take control over change. They decided what they wanted to change and what they wanted to preserve."

Things worked out differently than they had originally envisioned, because the villagers showed little interest in planning. "Development as it proceeds in these places is not like it is in the textbooks," explains Schubert. "In a traditional culture the concept of planning is totally alien. Their way of life is very different, very much day-to-day. They are very in touch with their families, their community, and the land. Life unfolds just fine for them without planning.

"We struggled with what to do," Schubert remembers, "but we finally decided that the program was inappropriate. We weren't making any headway in the areas where we were supposed to be. We went through a lot of soul searching, wondering, 'Do we have a role here?'

"We finally decided to begin in our own houses, to implement some of the things we thought they would be interested in." This included items such as sanitary latrines and wood-burning stoves. The stoves were a hit with local women, who "became very interested in building stoves together. They built and installed quite a few. This appealed to them because it got the smoke out of the kitchens and out of their eyes and lungs. The

continued from page 22

Jaye has been involved in exciting Peace Corps work at the Bururi Forest Reserve in East Africa. When he first arrived there, he served as temporary head of the Forest Reserve and was responsible for planning all work activities, training staff, and responding to the public's questions. Then he "focused on setting up a system of reserve surveillance, developing brochures to promote the reserve, creating topographical maps, and designing and building tourist hiking trails," says Mark. Jaye then took on the monumental task of "preserving the last national rain forest." This work involved transporting native animals and other species back to the region.

stoves also had the advantage of conserving wood. The women were also thrilled because stoves kept their pots clean and meant they had to gather less firewood."

Getting a Career Edge

Volunteers who complete natural resources assignments gain numerous advantages when seeking further employment. Conservation organizations, particularly those with international interests, such as the World Wildlife Fund, have a high regard for Peace Corps volunteers' work experience. Federal and state natural resource agencies also value the Corps experience. Many returned volunteers have found employment with the National Park Service or the Forest Service, often taking advantage of the "noncompetitive eligibility" (NCE) they receive to obtain federal government jobs.

Under NCE, returned volunteers may be appointed to federal positions without competing with the general public. If a candidate meets minimum requirements for a specific job, the hiring agency may hire the person directly, regardless of whether he or she is the most qualified candidate. Returned Peace Corps volunteers are eligible for the program for one year after they return home.

Federal agencies frequently seek out returned volunteers for employment. "My supervisor really likes NCE because it makes it easier and faster for him to fill a position—it makes it easier both ways," notes Tim Casten, who served in Guatemala as an agricultural extension volunteer from 1983 to 1985 before obtaining a position with the hazardous waste enforcement division of the Environmental Protection Agency.

NCE is a great opportunity for returned volunteers to get into the Forest Service, U.S. Fish & Wildlife Service, or National Park Service. These agencies all have a lot of returned volunteers working in the ranks, especially in their international divisions.

Many returned volunteers credit NCE with helping them cut through the process to find a job quickly. When Ruth Miller, a returned volunteer from Guatemala, applied through NCE, she had a position with EPA within three weeks. "I was in the right place at the right time," says Miller, now an environmental protection specialist for the agency's Policy Planning and Evaluation Office. EPA likes to hire returned volunteers, Miller believes, "because (1) volunteers have an altruistic attitude, they're optimistic, and they believe things can happen; (2) they have hands-on experience so they know things can happen, and (3) . . . they're very trainable and adaptable." EPA is full of returned volunteers, she continues, "You shake the tree and a whole bunch of us fall out."

Miller finds her experience as a forester in Guatemala (1982–85) helpful in her current job: "My Peace Corps experience made me extremely self-reliant and led me to be more creative and to realize a greater number of possibilities in terms of how to accomplish things. It taught me self-confidence, self-reliance, and patience."

Dave Reynolds served as a volunteer in Sierra Leone and the Ivory Coast from 1975 to 1977, working in planning and wildlife biology. Upon returning to the United States in 1977, he went straight into the National Park Service through NCE. "I found a tailor-made position, requiring overseas experience and proficiency in Spanish or French," he explains.

Reynolds believes Peace Corps experience is valuable in preparing a person for a position with the Park Service. "The biggest advantage is that volunteers gain two years of experience actually working in national parks—doing things they would never get to do at their age and grade in the United States. In some places, such as Botswana, people in their midtwenties are working as superintendents of parks larger than Yosemite with more wildlife. That looks pretty good on a résumé."

Although many returned volunteers report success through NCE, it must be noted that not everyone finds a job, and those who do, do not usually find them as quickly as Miller and Casten; nor are positions guaranteed. Whether or not they got their federal jobs through NCE, however, many returned volunteers report their Peace Corps experiences were helpful in obtaining their jobs or otherwise advancing their careers.

Pat Durst served in the Philippines from 1978 to 1980, part of the Corps' first agroforestry extension program. He worked with farmers on a variety of soil conservation measures and fuel wood programs. He is now the coordinator for Asia, Near East, and Europe for the Forestry Support Program, a joint project of the U.S. Department of Agriculture, Forest Service, and the U.S. Agency for International Development to provide support for USAID's forestry and natural resources programs.

Durst says his Peace Corps experience "allowed me to compete for and win a graduate assistantship position at North Carolina State with a professor interested in international forestry. I also applied for and received a Fulbright grant to return to the Philippines. I never would have received it without my prior experience in the country."

Durst believes his experience has helped in his current career. "Recruiters in these types of jobs value and almost require experience in field-level work—that's what needs to be done to make progress in tropical forestry. In the Philippines, everything was field-level: working with farmers, students, and community groups." Indeed, his program bears out this contention—six of seven permanent staff members are returned Corps volunteers.

Applying to the Peace Corps

For more information or an application, call the Peace Corps at (800) 424-8580 and ask for extension 916. Or, talk to former volunteers at one of the sixteen Peace Corps recruitment offices around the country. (Check your phone book's government listings.) You must be a U.S. citizen to apply. Almost all volunteers have at least a college degree. There is a limited number of assignments, and they are filled competitively. It may take six to nine months for an application to be processed.

ENVIRONMENTAL CAREER CENTER, INC.

The Environmental Career Center (ECC) is a nonprofit organization dedicated to helping people help the environment. ECC headquarters is located in Hampton, Virginia, with several regional offices currently planned for the Rocky Mountain and western regions. ECC assists college students, recent graduates, career changers, and other individuals to enter and develop their careers in the environmental field through the following integrated services:

> *Environmental apprenticeships*
> *Minority achievement program*
> *Environmental training*
> *Career counseling/job information*
> *Career conferences/seminars*
> *Environmental career research*
> *Publications*

ECC started as the Nature People career services in 1980 and created the first nationwide environmental and natural resources job bulletin. In 1988, the name was changed to the Environmental Career Center, and the headquarters moved from Wisconsin to Virginia to better assist job seekers.

"We are a small nonprofit with a big heart and a lot of in-the-trenches job-hunting experience," says John Esson, executive director and founder of ECC. Esson was on over one hundred federal registers in the 1980s and has sixteen years' environmental and natural resources experience working for federal and state agencies and private environmental consultants. (Federal registers are lists of qualified candidates for government jobs.) "I've been there, and we created the type of national environmental career service that can really help job seekers break into a career they'll love," Esson says.

ECC's new Environmental Partnership Program places aspiring environmental professionals in paid apprenticeships with federal agencies like EPA, state agencies, private industry, environmental consultants, and nonprofit environmental advocacy groups. Apprentices also attend environmental training sessions to further advance their careers.

The Environmental Career Center offers graduate-level environmental science and engineering courses at its Virginia office. ECC has joined with Colorado State University (CSU) to provide videotaped, accredited CSU graduate courses, including groundwater hydrology, environmental law, air quality control, and other courses for environmental professionals.

Career counseling and career seminars have been the heart of ECC's services over the years. ECC provides expert guidance on

- Successful job hunting strategies
- Current job openings nationwide
- Environmental career trends
- SF-171, state application, and résumé assistance

ECC publishes comprehensive federal salary surveys, updates on nontraditional environmental degree programs, and other key research reports on environmental careers. Staff are also frequent contributors to *Earth Work* magazine.

The Environmental Career Center welcomes inquiries from universities, organizations, and any individuals with a passion to work for environmental protection. For more information, contact ECC at

Environmental Career Center, 22 Research Drive, Suite 102, Hampton, VA 23666,
(804) 865-0605, fax (804) 865–0298

Chapter Three

GREEN GRADUATE SCHOOLS

Thinking about whether graduate school is worth the time and money? According to an expert with experience both in academia and in government hiring, "Graduate education is becoming the entry level degree in natural resources." Frank Gregg, former director of the U.S. Bureau of Land Management and of the University of Arizona's School of Renewable Natural Resources, says the days are gone when a bachelor's degree was enough for many natural resources jobs.

By contrast, jobs are readily available with a bachelor's in environmental engineering or in other environmental specialties, where demand exceeds supply. In the natural resources disciplines, however, there is a kind of buyer's market except in specialties such as wetlands ecology.

Several hundred colleges in the United States are churning out natural resources and environmental science graduates with advanced degrees. Out of 141 responses to a Conservation Fund survey of 390 colleges and universities with natural resources or environmental science programs, 37 percent had graduate programs. The survey did not include environmental engineering or waste management programs or agricultural specialties such as range science.

Because of the job satisfaction and contact with nature offered by natural resources careers, students are still attracted to them. According to a recent Society of American Foresters survey, enrollment (both graduate and undergraduate) at SAF-accredited schools rose 16 percent in 1991 compared with 1990. During the 1991–92 academic year, nineteen thousand students enrolled in these programs alone; however, fewer than a quarter were in graduate school. Ervin Zube, Gregg's colleague and a professor at Arizona's natural resources school, says that with this kind of ratio, "the agencies and consulting firms that have been hiring people can literally high-grade the market. They take the best people, and those are the people who have graduate degrees."

In environmental engineering and other technical fields, it is a seller's market, because tens of thousands of new engineers and other specialists will be needed by the year 2000. They will be able to find jobs easily with a bachelor's degree. However, because graduate degrees may double starting salaries in this field, engineers also have strong motivation to continue in school.

Another reason for the increasing emphasis on graduate education in both natural resources and environmental sciences is the information explosion that has occurred in the field over the past thirty years. The state of knowledge and the techniques available to natural resources professionals have improved dramatically. "It's naive to think we can produce a person in the same amount of time today with the breadth of knowledge needed to encompass the new science and the new tools," says Zube.

As interest in the environment has grown through the years, so have the number and kinds of programs in the natural resources and environmental fields. Today, the available curricula are almost as numerous and diverse as the schools that offer them. These programs may be found in forestry, natural resources, engineering, or agriculture schools, to name a few. Possible degree fields are even more numerous, including natural resources management, wildlife biology, forestry, parks and recreation, marine biology, coastal zone management, environmental law, environmental engineering, environmental studies, fisheries, parks and recreation, range science, fish and wildlife management, and environmental communications.

Curricula in the graduate schools show many stages of gradation between two educational philosophies—intensive scientific study in one discipline, on the one hand, and the interdisciplinary approach, on the other. A number of natural resource programs date back to the early 1900s. Created to focus on forestry, fisheries, wildlife, and agricultural issues in their regions, these schools often retain more traditional curricula. Environmental science programs founded in the 1960s and in the heady 1970s after the first Earth Day focused on the needs of the biophysical sciences and technology developments in addressing environmental problems. As the '70s progressed,

Knauss Sea Grant Fellow— Carol Collinson Kahl

While growing up Carol Collinson Kahl enjoyed Lake Michigan and Florida's beaches and envisioned herself as a female Jacques Cousteau.

But even though she is a marine science student now working at National Oceanic and Atmospheric Administration (NOAA), she probably will not be the counterpart of the famous French aquanaut. "Eventually I learned that long stints on research vessels were not my style," she says.

Collinson Kahl, thirty-six, grew up in Midland, Michigan. She received an associate of science degree in applied marine biology and oceanography from Southern Maine Vocational Tech and then her B.S. in education from the University of Southern Maine. She then worked at Bigelow Laboratories for Ocean Sciences in Boothbay Harbor, Maine, as a marine technician for a chemical oceanographer and also worked part-time at the Gulf of Maine Aquarium, in Saco, as an educational specialist.

With an urge to learn even more about marine science and policy, she enrolled in the Marine Environmental and

continued

environmental studies programs were born that were interdisciplinary from the start, combining biological and physical sciences with social sciences and humanities to resolve environmental problems.

continued from page 29

Estuarine Studies Program at the University of Maryland. The program's interdisciplinary approach, combined with the school's proximity to the federal government, made it an especially attractive choice.

Collinson Kahl is now one of only twenty-five Knauss Sea Grant fellows in marine science and policy and has been chosen to work in a damage assessment restoration program at the National Marine Fisheries Service Restoration Center in NOAA's Silver Spring, Maryland, office.

The Knauss Sea Grant Fellowship is a highly competitive and prestigious program, funded through the National Sea Grant College Program.

As a fellow, Kahl receives a $24,000 stipend plus some travel expenses for her year in Washington. Collinson Kahl is very pleased to have been chosen as a Knauss fellow. "The Knauss Fellowship is excellent. It gives me experience and exposure and a network of twenty-four other fellows in marine policy positions around Washington. I would encourage students to take advantage of any such internships or fellowship programs."

Although Collinson Kahl will probably be writing her thesis on oil-spill policy, she hopes to

continued

Academics differ as to which is the better path for today's graduate student. Gregg argues that many problems faced by today's natural resources professionals demand solutions combining elements of the physical sciences, social sciences, law, economics, and other disciplines. He points out that "the war in professional education in natural resources is always between traditionalists who want to turn out people whose education is fundamentally reductionist, who know quite a lot about a very narrow range of knowledge, and those who want to turn out people who are capable of finding relevance in apparently unrelated subjects."

One common argument in favor of the more traditional approach is the concern that if you educate people broadly enough to be really knowledgeable in all those fields, they will not have enough substantive understanding in one discipline to deal with technical issues. According to Zube, "I've come to the conclusion that an interdisciplinarian without a discipline has a hard time finding a home."

He argues, for example, that if the student gains a broad understanding of the field as an undergraduate and then zeros in at the graduate level on a specific discipline he or she will have gained the advantages of each style of education. The order can be reversed; a student educated in a traditional forestry undergraduate program would be advised to go into an interdisciplinary graduate program.

Stephen Kellert, professor at Yale's School of Forestry and Environmental Studies, points out the need for curricula that prepare students for policy making as well as communication outside their disciplines. He says most master's students at Yale are more interested in "policy making, decision making, management, and leadership types of positions rather than becoming, for example, forest technicians or range technicians."

James E. Crowfoot, professor and former dean of the University of Michigan's large and highly respected School of Natural Resources, examines the national gap between curricula and the severity of environmental problems in *Voices from the Environmental Movement*, by The Conservation Fund (see Book List). He predicts that "the pressure for interdisciplinary environmental problem solving in both curricula and research programs will continue to increase" because of the severity of environmental problems, even though teachers and students at many schools have not been trained to work in this way. "Problems such as hazardous and solid waste management, atmospheric deposition, ecosystem management, and forest planning cannot be adequately understood without the expertise of several disciplines."

The graduate school profiles and the list of colleges and universities that follow are intended to provide an overview of the types and breadth of programs available in each region of the country and are not meant as a ranking of schools. (The list does not include all disciplines or all schools for each discipline.)

If you cannot go to graduate school right away, don't panic. The Conservation Fund survey shows that most graduates in natural resources find employment in their field—80 percent of those with bachelor's degrees and 89 percent with advanced degrees. In the midst of a recession, those are high figures for any profession. If you are able to invest in graduate school, however, now is a good time.

continued from page 30

work on damage cases that NOAA handles and to help develop restoration plans for coastal areas. She will be working on the federal Coastal Wetlands Protection, Planning, and Restoration Act to halt degradation of Louisiana wetlands. She hoped to graduate in May 1993 with an M.S. in marine estuarine and environmental science.

A piece of advice she offers to students: Work with a mentor, either a teacher or a professional, to help guide you through career and educational choices.

Finally, she urges all aspirants to pursue their goals, no matter what their background or age. "I began graduate school when I was thirty-three—after ten years in the working world and a divorce."

GRADUATE SCHOOL PROFILES

DUKE UNIVERSITY'S SCHOOL OF THE ENVIRON-MENT stresses an interdisciplinary, systems approach to the study of natural resources and the environment. Students are encouraged to take full advantage of the university's resources by taking classes in business, economics, statistics, engineering, law, public policy, city and regional planning, conservation history, geology, and other areas. Programs focus on five areas: forest resource management, resource ecology, ecotoxicology and environmental chemistry, water and air resources, and resource economics and policy.

The school emphasizes its graduate professional program. "A large portion of our students are professional degree students: master of forestry or master of environmental management," says Mary Matthews, director of public relations. However, Duke also offers the traditional master of science and Ph.D. degrees.

The School of the Environment is organized around "centers" rather than traditionally structured departments. These centers are essentially umbrellas under which faculty can gather to do research addressing a specific problem. There are four formal centers: the

Wetlands Center, the Policy Center, the Marine Biological Center, and the Center for Tropical Conservation.

Students seeking admittance to the school have a wide range of educational backgrounds, according to Matthews.

THE ANTIOCH NEW ENGLAND GRADUATE SCHOOL has an across-the-board commitment to increasing environmental awareness. Mitchell Thomashow, co-chairperson of the Department of Environmental Studies, says, "Three years ago we started a program in which we are retraining faculty who are not environmental specialists so they can revise their courses and incorporate environmental . . . ideas into the normal teaching of everything from science to social sciences, engineering, business, and law." According to Thomashow, it is not specialists who are needed to solve environmental problems. "Environmentalists must be well-rounded, multiple-skilled leaders. They must be committed educators, managers, naturalists, philosophers, scientists, planners, and advocates," he says.

Two programs are offered, one leading to the master of science degree in environmental studies (ES) and the other leading to the master of science degree in resource management and administration (RMA). Both programs require certain core courses. Each student then individualizes his or her program of study by adding additional courses, practicum experiences, professional seminars, and electives.

The RMA program provides intensive training in natural resources policy, environmental science, and organization and management. Students are prepared for leadership roles in organizations involved in the management of natural resources. These include consulting firms, government agencies, businesses, and regional planning offices. Students focus on one of three core areas in the program—land use, water quality, or hazardous materials management.

The ES program trains students interested in a career in the environmental field. Students choose from among three concentrations: communications, administration, and education. The communications program trains people interested in careers as naturalists, curriculum developers, media consultants, and program directors. The administration program prepares students for management positions with environmental organizations and programs. Students concentrating in education who wish to teach can obtain certification in biology and general science, as well as elementary education.

UNIVERSITY OF MICHIGAN'S SCHOOL OF NATURAL RESOURCES allows students to choose to pursue a master's degree in one of three broad areas: resource ecology and management, environmental policy and behavior, or landscape architecture.

Environmental policy and behavior is geared toward the traditional social sciences. Students pursue such specialties as environmental policy, resource administration, environmental education, environmental psychology, and conservation behavior. This section contains the majority of the school's natural resources students, which is not surprising because the university boasts one of the nation's top policy schools.

Resource ecology and management is home to the more scientific disciplines. Students in this section pursue areas of study such as fisheries and aquatic ecology, conservation biology, wildlife ecology, and remote sensing.

The landscape architecture program is more straightforward. Individuals who complete the program graduate as certified landscape architects.

Students are typically in their late twenties and come from an environmentally related job with several years' experience. People entering the policy and behavior area come from a wide range of academic backgrounds, according to Sandy Gregerman, director of academic programs for the School of Natural Resources. "It's a very diverse student body—they don't necessarily have undergraduate degrees in environmental sciences or natural resources," she says.

Those in resource science and management usually have a more traditional science background in areas such as biology or ecology. All three areas feature an interdisciplinary focus. Students in each are required to take courses in resource policy, perhaps resource economics, land use planning, and at least a minimal amount of ecology in the behavior and policy area. Students in the resource and ecology area must include at least a minimum amount of policy and economics as part of their degree program.

Competition is keen for a place in the program. Last year more than five hundred people applied for admission. Only eighty were accepted.

THE UNIVERSITY OF ARIZONA'S SCHOOL OF RENEWABLE NATURAL RESOURCES offers degrees in four traditional programs. These are wildlife and fishery science, range science, watershed science, and landscape architecture. All these programs, except the last, offer a Ph.D. Students must have a background in renewable natural resources.

The program produces watershed hydrologists, range scientists, landscape architects, wildlife biologists, fisheries scientists, and other technical professionals. Also offered is an interdisciplinary renewable natural resources degree program that allows students to design their own curriculum based on the other four. The program requires students to complete graduate-level course work in at least two areas within the school and to complete a minor that is usually conducted outside the school.

Mary Soltero, graduate coordinator for the School of Natural Resources, believes students with an interdisciplinary degree are more employable: "The broader renewable natural resource studies give you a broader background in which you are not trained technically in just one area. You wouldn't be limited to just wildlife or watershed, but instead would have some understanding of all the natural resources."

Most graduates go to work for state or federal agencies, legislative budget offices, private consulting firms, or trade associations. Among the approximately 150 students enrolled in the school this past year were representatives of thirty-nine foreign countries. Many of these people return to their homelands to become the heads of government bureaus, according to Soltero.

Competition for admittance to this highly regarded school can be very tough. "The interdisciplinary program draws many outstanding candidates—many score in the ninety-ninth percentile on their GREs. We evaluate on a competitive basis and select only the best," says Soltero. Students entering the program must have an undergraduate science background.

TUFTS UNIVERSITY offers several graduate degree programs within an interdisciplinary framework. Most of the programs deal with pollution problems.

The environmental engineering master's program addressees pollution, from how it affects human health to how to design technologies to cope with and prevent pollution. The master's program in hazardous materials management is concerned with "everything from the use of gasoline and the storage of chlorine to the transportation and use of waste materials," according to Anthony Cortese, dean of environmental programs. Graduate programs in hazardous waste management are rare.

Tufts also offers master's programs in environmental policy and environmental health. Tufts's emphasis on pollution problems stems from a desire to "integrate the way we think about environmental issues to include natural resources and pollution that can affect human health," says Cortese. He believes the interdisciplinary approach is necessary because "we don't think about the environment as an integrated whole. Most people with expertise in natural resources management cannot deal very well with pollution-related issues except as they affect their resources." Conversely, "the people who study how to deal with pollution really know nothing about natural resources management."

New students come from a range of backgrounds, but most are professionals who have been out of school for five to eight years, Cortese estimates.

THE UNIVERSITY OF CALIFORNIA AT BERKELEY'S COLLEGE OF NATURAL RESOURCES is contained within the School of Forestry and Resource Management. It is home to programs in wildlife, range science, traditional or interdisciplinary forestry, soil and water sciences, and plant pathology, and a high-tech plant biology program, among others. It also houses less common programs, such as agricultural and natural resources economics.

Students can pursue traditional courses of study. "If you're in plant biology and your goal is to improve environmental management by manipulating genes, you go through that program; you're in a hard-core science program," according to Sally Fairfax, associate dean of the College of Natural Resources. The forestry program is home to most of the environmental management students and is "wildly interdisciplinary," she says. Fairfax cites a number of students pursuing unusual interdisciplinary courses of study—for example, issues affecting women and development in third world countries, the culture and management of pastoral lifestyle and ecosystem interfaces, or medicinal herbs.

Students often pursue joint programs with other departments. "If you want a degree in environmental law, you can come here and get a master's degree in resource management and a J.D. done jointly," explains Fairfax. More typically, such programs are done between the College of Natural Resources and the Schools of Public Policy or Environmental Design.

Whichever path the student chooses, she or he will find wide-open opportunities for interdisciplinary study. "I explain that it's like a bazaar in a big tent," says Fairfax. "You have to find the most felicitous way to get your nose in under the tent. Once you do, you can wander from booth to booth and take what you want as long as you maintain the support of an advisory committee and are headed toward a reasonable dissertation topic."

Fairfax points out Berkeley's strength as a major research center and its strong schools of law and business administration (among others) that make the university an attractive alternative for interdisciplinary study.

YALE UNIVERSITY'S SCHOOL OF FORESTRY AND ENVIRONMENTAL STUDIES is strictly a graduate professional school. Master's degrees are offered in forestry, forest sciences, and environmental studies. These are conventional two-year degrees. There are also five joint-degree programs offered in coordination with the Schools of Organization and Management, Law, Epidemiology, and Public Health. They may include additional course work in areas such as international development, economics, or international relations. Two doctoral degrees are offered, a Ph.D. and a doctor of forestry and environmental studies.

The Yale program covers a wide range of resource specialties—just about everything except range and marine science, according to Dr. Stephen Kellert, associate dean of forestry and environmental studies. Students tend to focus on one of three basic areas: ecosystem ecology, resource management (management technology), and resource policy. Individuals typically choose a subarea of concentration within these areas, such as water, wildlife, or a particular type of problem, such as land use management.

Those individuals completing a master's degree tend to be more interdisciplinary and generalist and less technical in their orientation, according to Kellert: "If you want technical training—say, you want to be a technical forester—I'm not sure Yale is the best place for you to go." Kellert says most master's students at Yale generally are more interested in policy-making and management positions than in technical scientific jobs. Some doctoral candidates at Yale do focus on a particular area of resource science or management.

Students enter the program from a variety of undergraduate backgrounds. However, "the majority have a science background of some sort. If you don't have a fair amount of scientific course work, it's very hard to get into the school because it's so competitive," says Kellert.

Although some students enter the program directly from undergraduate study, the typical new admission has three to five years of professional experience, often in the environmental field. In admission decisions, "we pay a lot of attention to the kinds of practical experience people have had. Some may not have had undergraduate training in the sciences, but their practical experience in the environmental field tends to qualify them for consideration, at least. If they have very good experience and accomplishments, we'll typically admit them on the condition they take certain kinds of scientific courses," Kellert explained.

THE UNIVERSITY OF WISCONSIN AT MADISON offers graduate instruction through the School of Natural Resources, located within the College of Agriculture, and the Institute for Environmental Sciences. Both programs are interdisciplinary.

A student wishing to pursue a graduate degree within the School of Natural Resources first enrolls with a declared area of interest. He or she then works with a faculty committee to define the course of study and decide which department will be the student's base. "If a student is interested in emphasizing wildlife or forestry and wants to do an interdisciplinary master's degree, he or she still has to identify a department through which to work, such as agricultural economics or rural sociology," explains Don Field, associate dean of the College of Agriculture and director of the School of Natural Resources. "We insist that students who are going to be housed in a forestry program should at least understand forestry and the way foresters look at the world."

The School of Natural Resources includes forestry, wildlife, landscape architecture, environmental sociology, natural resources economics, communications, recreation, and environmental education.

The other unit on campus is the Institute for Environmental Studies (IES), which offers an interdisciplinary degree in a number of program areas, including water resources and land resources. IES is a freestanding unit within the university made up of faculty from the Colleges of Engineering, Law, Agriculture, Life Sciences, and Family Resources. "The difference between this program and that of the School of Natural Resources is that a natural resources student interested, for example, in studying nature from a wildlife perspective would have to take a core group of courses in wildlife making up 25–50 percent of his or her graduate program. A student wanting to be more eclectic and less disciplinary might prefer IES," explains Field.

THE UNIVERSITY OF IDAHO COLLEGE OF FORESTRY, WILDLIFE, AND RANGE SCIENCES began in 1909 as a department in the College of Agriculture, and it was considered an experiment. Today, more than sixty faculty members guide the academic careers of more than 400 undergraduate students and 180 graduate students—more than 25 percent of them women. Students pursue B.S., M.F., M.S., and Ph.D. degrees, selecting courses of study from a myriad of options within the departments of forest resources, forest products, fish and wildlife resources, range resources, and resource recreation and tourism.

The College of Forestry, Wildlife, and Range Sciences is spread out over about five hundred miles of the scenic landscape to the north and south of Moscow, Idaho, and encompasses the Clark Fork Field Campus; the Lee A. Sharp Experimental Area; the McCall Field Campus, home of the college's summer camp (now Wildland Ecology); and the Taylor Ranch Wilderness Field Station, where FWR scientists and students pursue research possible only in a wilderness ecosystem. Students participate in other ranch activities through wilderness research internships, summer positions for which students compete. For instance, at the university's 7,300-acre experimental forest, members of the student logging crew—paid positions—work in all phases of logging. Another student organization, the Student Management Unit, controls a parcel of the experimental forest, working with faculty advisers to work out—and carry out—appropriate management techniques. And in the process, the students receive invaluable "real-world" experience.

The college's educational philosophy is firmly established on the principle of integrated resource management, on breaching the barriers among disciplines with the simple proposition that all natural resource professionals share the same ecosystem. To reinforce this proposition, all departmental curricula share particular courses; all students are encouraged to explore disciplines other than their own. Idaho's innovative approach to forest management called "adaptive forestry" stresses management that embraces all forest uses and balances all its values, whether commodity, environmental, aesthetic, or recreational.

FWR faculty members are currently engaged in over two hundred research projects—topics ranging from the strategies of wolf recovery and the management of grizzly bear habitat to wood-based alternative energy sources, from predicting the risk of wilderness wildfires with computer expert systems to assessing the role of resource-based tourism in rural communities. Students are involved in virtually all current projects.

The University of Idaho College of Forestry, Wildlife, and Range Sciences currently offers over thirty scholarships.

NATURAL RESOURCES AND ENVIRONMENTAL GRADUATE SCHOOLS

Listed alphabetically are representative graduate schools in the natural resources and environmental disciplines. This list is not all-inclusive; nor is it a ranking of schools. For each school, disciplines listed in the key are shown by number. In some cases the name of only one department is given. In addressing correspondence, indicate the name of the department in which you are most interested, or contact the office of graduate admissions. Space prohibited listing all the schools for each discipline, and the list does not include other relevant disciplines, such as other biological sciences (including, for example, zoology), earth and atmospheric sciences, resource economics, energy management and policy, environmental education, or environmental journalism.

KEY TO PROGRAMS AT EACH SCHOOL

1: Wildlife and fish science (*=wildlife only)
2: Range management/sciences
3: Parks and/or recreation
4: Environmental engineering
5: Landscape architecture
6: Environmental studies
7: Natural resources management/conservation
8: Forestry
9: Wildlife management

10: Fisheries management
11: Oceanography/marine science
12: Environmental science
13: Hazardous waste management
14: Environmental law

Alaska, University of, School of Wildlife and Fisheries, Fairbanks, AK 99775-0990,
    (907) 474-7671, **4,9,10,11**

Antioch University, New England Graduate School, 103 Roxbury Street, Keene, NH 03431,
    (603) 357-3122, **6**

Arizona, University of, School of Renewable Natural Resources, Tucson, AZ 85721,
    (602) 621-7255, **1,2,5,7**

Auburn University, School of Forestry, Auburn, AL 36849, (205) 844-1007, **1,8,9,10**

California, University of, School of Forestry and Resource Management, Berkeley, CA 94270,
    (510) 642-0376, **2,4,7,8,9,14**

California, University of, at San Diego, Scripps Institute of Oceanography, La Jolla, CA 92037,
    (619) 534-4831, **11**

California Institute of Technology, Pasadena, CA 91125, (818) 356-6341, **4**

Cincinnati, University of, School of Engineering, Cincinnati, OH, (513) 556-2946, **4**

Clemson University, Clemson, SC 29634-0919, (803) 656-5572/3195, **1,3,4,8**

Colorado State University, Fort Collins, CO 80523, (303) 491-1101, **1,2,3,4,8,9,10**

Connecticut, University of, Department of Natural Resources Management and Engineering,
    Storrs, CT 06269-4087, (203) 486-2840, **4,7,9,10**

Cornell, University of, Department of Natural Resources, Ithaca, NY 14853, (607) 255-2298,
    1,4,5,8,11

Duke University, School of the Environment, Durham, NC 27706, (919) 684-2135, **7,8,12**;
    in Beaufort, NC 28516, **11**

Florida, University of, Gainesville, FL 32611, (904) 392-1791, **1,2,4,8,9,10**

Georgia, University of, Athens, GA 30602, (404) 542-2968, **7,8,9,10,11**

Georgia Institute of Technology, Atlanta, GA 30332, (404) 894-4154, **4**

George Washington University, 37th and O Streets, NW, Washington, DC 20057,
    (202) 625-3051, **14**

KEY TO PROGRAMS AT EACH SCHOOL		
1: Wildlife and fish science (*=wildlife only)	**5:** Landscape architecture	**10:** Fisheries management
2: Range management/sciences	**6:** Environmental studies	**11:** Oceanography/marine science
3: Parks and/or recreation	**7:** Natural resources management/conservation	**12:** Environmental science
4: Environmental engineering	**8:** Forestry	**13:** Hazardous waste management
	9: Wildlife management	**14:** Environmental law

Hofstra University, 1000 Fulton Street, Hempstead, NY 11550, (516) 560-6700, **14**

Humboldt State University, College of Natural Resources, Arcata, CA 95521, (707) 826-3256, **1,4,7,8,9,10,12**

Idaho, University of, College of Forestry, Wildlife, and Range Sciences, Moscow, ID 83843, (208) 885-6441, **1,2,3,8,9***

Indiana, University of, School of Health, Physical Education, and Recreation, Bloomington, IN 47405, (812) 855-4711, **3**

Iowa State University, Ames, Iowa 50011, (515) 294-1166, **5,8,10**

Kansas State University, Manhattan, KS 66506, (913) 532-6011, **1,2,5**

Lewis and Clark University, 10015 SW Terwillinger Boulevard, Portland, OR 97219, **14**

Louisiana State University, Baton Rouge, LA 70803, (504) 388-4131, **1,8,9,10**

Maine, University of, Orono, ME 04469, (207) 581-1110, **3,8,9,14**

Maryland, University of, Frostburg, MD 21532, **1,9,10**; *College Park, MD 20742,* **11,14**

Massachusetts, University of, Amherst, MA 01003, (413) 545-2665, **1,4,5,7,8,10**

Miami, University of, Rosenstiel School of Marine Sciences, 4600 Rickenbacker Causeway, Miami, FL 33149, **11**

Michigan State University, College of Agriculture and Natural Resources, East Lansing, MI 48824, (517) 355-1855, **1,3,8,9,10**

Michigan, University of, School of Natural Resources, Ann Arbor, MI 48109-1115, (313) 364-2550, **5,6,7,8,10,11,14**

Minnesota, University of, St. Paul, MN 55108, (612) 624-1234, **3,8,9,10**

Mississippi State University, Drawer FR, Mississippi State, MS 39762, (601) 325-2952, **3,8,9,10**

Missouri, University of, School of Forestry, Fisheries, and Wildlife, Columbia, MO 65211, (314) 882-6446, **1,3,8,9,10**

Montana, The University of, Missoula, MT 59812, (406) 243-5221, **2,3,6,8,9**

Montana State University, Department of Biology, Bozeman, MT 59717, (406) 994-4548, **1,9**

Nebraska, University of, Institute of Agriculture and Natural Resources, Lincoln, NE 68583, (402) 472-7211, **2,8,9,10**

Nevada, University of, 1000 Valley Rd., Reno, NV 89512, (702) 784-4000, **2,8**

KEY TO PROGRAMS AT EACH SCHOOL		
1: Wildlife and fish science (*=wildlife only)	5: Landscape architecture	10: Fisheries management
2: Range management/sciences	6: Environmental studies	11: Oceanography/marine science
3: Parks and/or recreation	7: Natural resources management/conservation	12: Environmental science
4: Environmental engineering	8: Forestry	13: Hazardous waste management
	9: Wildlife management	14: Environmental law

New Jersey Institute of Technology, Graduate Admissions, Newark, NJ 07102, **13**

New Mexico State University, Fishery and Wildlife Sciences, Box 30003, Dept. 4901, Las Cruces, NM 88003-0003, (505) 646-1544, **1,2,10**

North Carolina State University, Box 7103, Raleigh, NC 27695-7103, (919) 737-2437, **1,3,5,8**

North Carolina, University of, Chapel Hill, NC, **3,4,11,12**

Nova University, Oceanographic Center, 8000 N. Ocean Dr., Dania, FL 33304, **11**

Ohio State University, 2021 Coffey Road, Columbus, OH 43210, (614) 292-2265, **1*,5,7**

Oklahoma State University, Stillwater, OK 74078, (405) 744-5000, **1,2,8,12**

Oregon State University, Corvallis, OR 97331, (503) 737-0123, **1,2,3,8,10,11**

Oregon, University of, Eugene, OR 97403, (503) 686-3201, **14**

Pace University, 78 North Broadway, White Plains, NY 10603, (914) 682-7000, **14**

Pennsylvania State University, University Park, PA 16802, (814) 863-1663, **3,4,8,9**

Purdue University, Department of Forestry and Natural Resources, 1159 Forestry Building, West Lafayette, IN 47907-1159, (317) 494-3590, **1,3,8,9**

Rhode Island, University of, Graduate School of Oceanography, Narragansett Bay Campus, Narragansett, RI 02882, **11**

Rutgers University, Cook College, New Brunswick, NJ 08903, (201) 932-9236, **1*,6,8,14**

San Francisco, University of, Environmental Management Program, 2130 Fulton Street, San Francisco, CA 94117, Interdisciplinary includes **12,13**

Southern Illinois University, Carbondale, IL 62901, (618) 453-2121, **1,3,8**

State University of New York, College of Environmental Science and Forestry, One Forestry Drive, Syracuse, NY 13210-2778, **1,4,5,8,9,10,12**

Tennessee, University of, Knoxville, TN 37901, (615) 974-7126, **1,8**

Texas A&M University, College Station, TX 77843, (409) 845-3431, **1,2,8,9**

Texas, University of, Austin, Texas, **4**

Tufts University, Medford, MA 02155, (617) 627-3211, **4,6,13**

Tulane University, 6823 St. Charles Avenue, New Orleans, LA 70118, (504) 865-5731, **14**

Utah State University, Logan, UT 84322, (801) 750-2445, **1,2,3,4,5,7,8,10**

KEY TO PROGRAMS AT EACH SCHOOL		
1: Wildlife and fish science (*=wildlife only)	**5**: Landscape architecture	**10**: Fisheries management
2: Range management/sciences	**6**: Environmental studies	**11**: Oceanography/marine science
3: Parks and/or recreation	**7**: Natural resources management/conservation	**12**: Environmental science
4: Environmental engineering	**8**: Forestry	**13**: Hazardous waste management
	9: Wildlife management	**14**: Environmental law

Utah, University of, Salt Lake City, UT 84112, (801) 581-7281, **14**

Vermont, University of, School of Natural Resources, School of Law, Burlington, VT 05405, (802) 656-4280, **1,6,7,8,14**

Virginia Polytech Institute and State University, Blacksburg, VA 24061-0324, (703) 231-5481, **1,3,4,8,10**

Virginia, University of, Charlottesville, VA 22903, **4,6,12**

Washington, University of, Seattle, WA 98195, (206) 543-6475, **3,5,8,9,10,11,14**

Washington State University, Pullman, WA 99164, (509) 335-4000, **2,3,7,8,9**

Wayne State University, Detroit, MI 48202, (313) 577-2876, **6,7,13**

West Virginia University, Morgantown, WV 26506, (304) 293-2941, **3,8,9,10**

Widener University, Chester, PA 19013, (215) 499-4126, **14**

Wisconsin, University of, at Madison, School of Natural Resources, Institute of Environmental Studies, Madison, WI 53706, (608) 262-6968, **1*,5,6,7,8**

Woods Hole Oceanographic Institution, Woods Hole, MA 02543, (degrees through MIT, Boston University,, University of Maryland, etc.), **11**

Wyoming, University of, Laramie, WY 82071, (307) 766-4227, **1,2,9,10**

Yale University, School of Forestry and Environmental Studies, 205 Prospect Street, New Haven, CT 06511, (203) 432-5100, **1,3,6,7,8,11,12,14**

SOURCES: Peterson's *College Guides;* National Wildlife Federation *Conservation Directory; Graduate Record Exam/CGS Directory of Graduate Programs in Natural Sciences;* Society of American Foresters; American Fisheries Society; National Recreation and Park Association; Environmental Career Center; Bureau of Land Management; *The Gourman Report; The Environmental Career Guide,* by Nicholas Basta (John Wiley, 1992); and *Earth Work* magazine, June 1992, list of graduate schools in marine science and oceanography.

KEY TO PROGRAMS AT EACH SCHOOL		
1: Wildlife and fish science (*=wildlife only)	5: Landscape architecture	10: Fisheries management
2: Range management/sciences	6: Environmental studies	11: Oceanography/marine science
3: Parks and/or recreation	7: Natural resources management/conservation	12: Environmental science
4: Environmental engineering	8: Forestry	13: Hazardous waste management
	9: Wildlife management	14: Environmental law

Chapter Four

SURVEYING THE FIELD: SALARIES

Traditionally people seeking conservation and environmental jobs have assumed that they would make less money than those in other fields. They were, however, more than willing to forsake monetary rewards for the promise of greater job satisfaction, the opportunity to serve a good cause, and/or the privilege of working in nature. The days when environmentalists worked mostly for love, not money, still persist at some nonprofits and even in coveted government positions such as that of park ranger. Some who start out in these jobs remain in them throughout their careers, while others choose different avenues because of the demands of family and home. Many nonprofits, in particular, have found themselves with few qualified midlevel employees.

The good news for the 1990s is that it is now possible to find a job that will fill your heart *and* your wallet. This chapter will help you compare salaries in your field of interest—or pick a field and a specific kind of job goal more likely to pay you materially as well as spiritually. *Earth Work* has made salary comparisons extrapolated from three sources: a survey of salaries in nonprofit organizations conducted by Jon Roush and the staff of Canyon Consulting; a survey of salaries for specific government jobs conducted by the Environmental Career Center; and figures on for-profit jobs by the Environmental Careers Organization.

This salary information will help you evaluate the upcoming chapters on jobs in the nonprofit, federal, and for-profit sectors. You can not only compare similar jobs across the sectors, but also determine which jobs and which agencies are most promising within a sector. For example, even though nonprofit salaries are generally lower, CEOs and development professionals may do better in these organizations than in government. And among government agencies, EPA offers the highest pay, while the Forest Service offers the most jobs. However, among the most highly paid individual positions surveyed is that of fish and wildlife administrator at the U.S. Fish and Wildlife Service.

NONPROFIT CONSERVATION ORGANIZATIONS

In the past, data on salaries in the nonprofit conservation field have been scanty. In an attempt to fill the information gap, Canyon Consulting, a national firm specializing in organizational and management planning for nonprofit environmental organizations, conducted salary surveys from 1990 to 1992. These invaluable surveys provide average salaries for eighteen position categories based on information from the organizations replying each year (77 to 111 groups). From this primary source and additional information, *Earth Work* has estimated salaries for 1994. These salaries are averages calculated from a broad range of groups and pay scales.

AVERAGE NONPROFIT SALARIES

CEO: $70,000
Vice President: $65,000
Financial Officer: $57,000
Science Director: $56,000
Membership Director: $48,000
Project Director: $47,000
Media Director: $47,000
Senior Field Staff: $37,000
Midlevel Professional: $28,000
Secretary: $20,000

The Canyon Consulting surveys did not include employees of universities or other learning centers who work in conservation-related jobs. Adjusting past data for inflation, it seems that university employees of all types (ranging from research aide to professor) with jobs related to the environment average about the same as senior staff at nonprofit membership organizations.

Several broad trends have held over the past five years for nonprofit jobs:

↝ Salaries at nonprofits range greatly, depending usually on the size of the organization. For example, a CEO's salary ranges from $25,000 at some organizations with three or fewer employees to more than $200,000 at several large organizations. The vice president's salary is almost as high as the CEO's because smaller groups do not have vice presidents, so the average reflects only salaries of larger companies, which tend to pay more.

- Nonprofit organizations are investing heavily in development. Salaries for development directors have risen faster than those in any other job category, including CEO, reflecting the need for increased fund-raising due to the economy and dwindling contributions to charities in general. The average salary for directors of development approaches $60,000, with many topping $100,000.

- The median salary for midlevel professionals is about half that of CEOs.

- Most job categories have not experienced significant increases or decreases, with the exceptions being increases in the salaries of presidents, development directors, and scientists with advanced degrees.

THE "BIG FIVE" FEDERAL ENVIRONMENTAL AGENCIES

Ask most environmental professionals to name the primary federal environmental management agencies, and they will usually say the U.S. Forest Service, National Park Service, Bureau of Land Management, Fish and Wildlife Service, and Environmental Protection Agency—the "big five." These agencies employ natural resource management and environmental compliance specialists to manage more than 630 million acres of public lands and enforce an ever-increasing number of complex environmental laws. The Environmental Career Center evaluated and compared salaries for natural resources and environmental permanent employees of the big five agencies for *Earth Work.*

Some federal agencies with significant environmental professional workforces were not included in this survey. They include (among others) the Department of Energy (DOE), Department of Defense (DOD), Bureau of Indian Affairs, National Marine Fisheries Service, NASA, Bureau of Reclamation, and Soil Conservation Service.

"What's your grade level?" or "What kind of grade levels does your office have?" are fairly common questions among federal employees. Grade levels are standard salaries found throughout the federal government. Professional and technical positions are part of the General Schedule (GS). Most college graduates obtain their first permanent federal position at the GS-4 to GS-9 level, for which starting salaries range from about $16,000 to $28,000, depending upon position and geographic area. Within each grade are step increases that are based on years of satisfactory service. The Environmental Career Center used a typical salary within each grade for calculations in its survey.

Administrative positions usually pay more, but they often take employees out of the field and place them behind the desk (some of these jobs are in the General Schedule, and others are under GM, the designation for administrative/management positions). Salaries may range from GS-11 (approximately $33,000–$46,000) to GS-15 ($66,000–$90,000) and up into the Senior Executive Service (ES)/Executive (EX) grades, which pay up to about $150,000 annually. Most administrators and senior managers are at GS-12 to GS-14, which range from about $40,000 to more than $75,000, depending on locality and step.

SURVEY METHODS

The Environmental Career Center sent a questionnaire to the federal agencies and asked for data on the total numbers of permanent employees in each job series in each grade. A job series is a specific position such as forester, wildlife biologist, or range conservationist. The center requested data on thirty-four specific job series, from the typical park ranger, forestry technician, or wildlife biologist to less common environmental or natural resource positions, such as land surveyor, cartographic technician, or special agent (wildlife).

The data represent a complete count of all permanent employees and constitute a sample survey. According to the survey, at the beginning of this decade there were almost thirty-seven thousand permanent natural resource and environmental employees. Figure 1 shows the distribution of employees for each job series surveyed. Forestry technician, forester, and park ranger are the most common positions in these agencies. Figure 2 shows average salaries by job series.

Unlike the other job series mentioned, the NPS park ranger series is not a professional one and has no positive educational requirement in the biological or other natural resources fields. It is included here because the park ranger positions entail natural resources management responsibilities and are commonly thought of as comparable jobs to the others discussed in this chapter.

WHICH AGENCIES PAY THE HIGHEST SALARIES?

Environmental Protection Agency (EPA) employees are the highest paid among the big five agencies (see Figure 2). EPA employees in the job series surveyed average more than $45,000 per year. EPA's environmental compliance mission requires more environmental scientists, environmental protection specialists, and engineers than the other agencies. The National Park Service, Bureau of Land Management (BLM), Forest Service, and Fish and Wildlife Service generally have more technician and field personnel slots, which pay less. National Park Service and Forest Service employees have the lowest average annual salaries, at little more than $30,000. U.S. Fish and Wildlife technical/professional employees average about $34,000 annually, and the average BLM employee is paid only slightly more, at an average of less than $36,000.

HOW DO FEDERAL JOBS COMPARE WITH JOBS IN THE PRIVATE SECTOR?

Industry and environmental engineering and/or consulting firms pay higher salaries than federal agencies surveyed and nonprofit organizations. For example, the Environmental

FIGURE 1. NUMBER OF EMPLOYEES EACH FEDERAL JOB SERIES

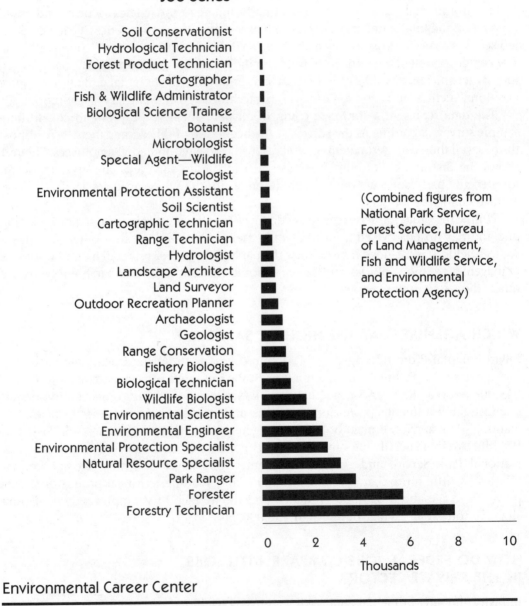

Job Series

(Combined figures from National Park Service, Forest Service, Bureau of Land Management, Fish and Wildlife Service, and Environmental Protection Agency)

Thousands

Environmental Career Center

FIGURE 2. AVERAGE FEDERAL SALARY BY JOB SERIES

Job Series with 500+ employees

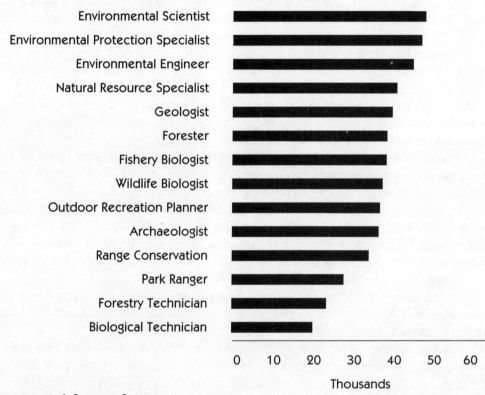

Environmental Scientist
Environmental Protection Specialist
Environmental Engineer
Natural Resource Specialist
Geologist
Forester
Fishery Biologist
Wildlife Biologist
Outdoor Recreation Planner
Archaeologist
Range Conservation
Park Ranger
Forestry Technician
Biological Technician

0 10 20 30 40 50 60

Thousands

Environmental Career Center

Career Center estimated that technical managers at consulting firms make 36.5 percent more than midlevel (GS-9/11) federal employees and 31.6 percent more than people in similar positions in nonprofit organizations. Overall, nonprofit salaries are very similar to federal salaries.

Working for Love and *Money*

As one would expect, more employees earning over $60,000 work for private companies than for the government or nonprofits. The salary level earned is affected by the type of for-profit company that employs an individual.

Based on information from the Environmental Careers Organization, it can be estimated that currently in manufacturing firms, most professional employees earn above $35,000 and perhaps a third make more than $60,000.

Environmental engineering and consulting firms pay top salaries. The highest percentage of individuals in environmental jobs earning over $80,000 (7 percent) work for these firms; perhaps half of the professional employees in these companies make $50,000 or above.

Salaries at laboratories and scientific research facilities vary widely, ranging from about $25,000 to more than $100,000. Many scientific aides and technicians earn at the lower end of the scale.

Managers earn higher salaries than nonmanagers in engineering, scientific/technical, and other positions. Engineers, for example, can peak at salaries of $60,000 in many companies unless they enter the management ranks.

Advanced degrees are in demand. With the need for higher skill levels and more multidisciplinary backgrounds, professionals are seeking master's and doctoral degrees to increase their understanding of increasingly complex issues.

The environmental career field is growing. Salaries at land management agencies such as the National Park Service, Forest Service, Fish and Wildlife Service, and BLM are lower than salaries for environmental compliance positions with EPA, in industry, and in consulting firms. However, lower salaries are balanced by the opportunity to live close to pristine areas in national parks, forests, and wildlife refuges.

Salaries should continue to rise moderately in the profit sector. New Clean Air Act Amendments (CAAA of 1990) should generate an estimated sixty thousand additional positions by the year 2000.

Although salaries in the nonprofit sector are more vulnerable than those in government agencies to economic fluctuations, the federal government is in the process of downsizing in some agencies. However, although certain categories will experience a decrease in numbers, others will increase.

Federal salaries should remain fairly stable to slightly increasing as federal environmental laws are implemented (e.g., CAAA of 1990, Clean Water Act, storm water runoff and wetlands definition).

As in other professions in which one tries to make a living while doing good work, you'll be working for both love and money. Despite economic ups and downs, however, you can be assured that the environmental field is on a long-term growth curve. You are needed!

Chapter Five

GOVERNMENT GREEN JOBS

The U.S. government has a wide variety of interesting and challenging career opportunities available to you in environmental protection and natural resources management. Work can vary from that of field technicians conducting studies of fish and wildlife and ecological surveys while enjoying the great outdoors to that of environmental program managers and administrators who plan, budget, and oversee critical environmental restoration programs. They work not only for the agencies you would expect—such as the Environmental Protection Agency or the Fish and Wildlife Service—but also for the Department of Energy (DOE), Department of Defense (DOD), and other agencies. Federal agencies need not only skilled scientists, engineers, and technicians but also management analysts, public relations specialists, environmental attorneys, project managers, and other individuals skilled in environmental and natural resources management.

The Environmental Career Center study for *Earth Work* featured in chapter 4 identified nearly thirty-seven thousand permanent environmental and natural resource positions with the "big five" federal environmental agencies—the Environmental Protection Agency, National Park Service, Fish and Wildlife Service, Bureau of Land Management, and Forest Service. These agencies attract the greatest interest and the greatest competition for jobs.

But other agencies provide additional opportunities for interesting and challenging careers in protecting our nation's environment. Some of these opportunities are at the National Marine Fisheries Service, Bureau of Reclamation, Bureau of Indian Affairs, Army Corps of Engineers, Department of Defense agencies, Department of Energy facilities and laboratories, the Federal Energy Regulatory Commission, and the new National Biological Survey.

In the past two decades in the United States, there has been nearly exponential growth in environmental laws and regulations leading to more stringent standards. This result of environmental awareness and advocacy has ultimately created more jobs responsible for planning, implementing, and enforcing environmental requirements.

Even though the federal government has sought to limit the number of civil service workers overall through reductions-in-force and closing and consolidating of facilities, the environmental workforce has grown. The greatest growth in federal environmental jobs occurred after the Resource Conservation and Recovery Act (RCRA) was reauthorized in 1984 and after the Superfund Amendments and Reauthorization Act (SARA) of 1986 and the Clean Air Act Amendments (CAAA) of 1990.

An EPA administrator of air quality programs estimates that the CAAA will add sixty thousand new air quality jobs in both private and public sectors. Part of this growth includes more research and regulatory positions at EPA and additional environmental protection specialist, scientist, and engineer positions at other agencies that must comply with the new requirements. New requirements of the Endangered Species Act and Clean Water Act should also increase the workload for both regulatory and regulated agencies and lead to more jobs in both public and private sectors.

The first, most critical, step to securing a federal environmental job is completing a well-crafted federal application form.

The "Hidden" Federal Job Market

Most entry-level environmental and natural resource vacancies filled through OPM have been for Department of Defense agencies. That indicates that there is somewhat less competition for these jobs than for jobs in the traditional natural resource agencies and perhaps a "hidden market" for their vacancies.

In early November 1993, the Army, Navy, Air Force, and Marine facilities had thirty environmental or natural resource vacancies open to any qualified outside candidate. At the same time, the "big five" federal agencies (EPA, NPS, USFWS, BLM, USFS) had a total of twenty-seven vacancies open to all candidates.

Now, which agencies do you think had the most applicants? Which agencies probably had the least number of applicants? Where are your odds for success greater? DOD offers excellent advancement potential, full federal benefits (including transfer eligibility to USFS, USFWS, NPS, etc.), good career training courses, travel, and even some opportunities for overseas assignments.

APPLICATION FORMS

The classic one- or two-page résumé may be the standard in the private sector, but the Standard Form 171 (SF-171) and Office of Personnel Management (OPM) Form 1170/17 (list of college courses) are the primary federal application forms. Résumés do not get serious attention for most positions.

The SF-171 is a long, tedious form that takes time to complete. Make that your advantage instead of a hindrance. Most other applicants, including current federal employees, do not keep their SF-171 updated and poised for the next job opening.

You must be ready to send your SF-171 immediately upon discovering the vacancy announcement. Many vacancies have short application periods, and you could miss an open window of opportunity if your SF-171 is not kept current.

Typing your application on SF-171 software is the best strategy for preparing and updating it. You can revise your SF-171 to match the wording of specific job screening criteria known as knowledges, skills, and abilities. It is easier for rating panels to score your experience, knowledge, awards, and self-improvement if they see similar language in both the SF-171 and the job vacancy announcement and rating criteria.

There is a Word Perfect template for SF-171 preparation. There is also a software package called TruForm that provides an easy-to-use SF-171 for the Macintosh. Other popular SF-171 software packages include

SF-171 Automated, Software Den, 103 Loudoun Street SW, Leesburg, VA 22075

Quick and Easy, Federal Research Service, P.O. Box 1059, Vienna, VA 22183-1059

Fedform—171 Laser, Arumon Group, P.O. 25090, Arlington, VA 22202

FEDERAL HIRING PROCEDURES

There is no easy way to obtain a permanent environmental or natural resources position in the federal government; competition is keen. There are at least nine routes to a permanent federal job, and it is best to seek a position through several at once.

1. *Stay-in-School program*

2. *Federal Junior Fellowship program*

3. *Cooperative education programs*

4. *Outstanding Scholars program*

5. *President's Management Intern program*

6. *Registers of qualified candidates*

7. *Case examining by U.S. Office of Personnel Management (OPM)*

8. *Direct agency hiring*

9. *Peace Corps alumni program*

STAY-IN-SCHOOL PROGRAM

The Stay-in-School program is designed to provide employment for high school or other postsecondary students who have financial need or have disabilities. Students must be at least 16 years old; maintain satisfactory grades as they work toward a diploma, certificate, or degree; and qualify under financial need criteria. Salary is based on qualifications, and students are eligible for promotions, awards, and other pay increases. Students also qualify for paid sick leave and annual leave (vacation), accruing at four hours each per pay period.

About twenty thousand students participate in the Stay-in-School program nationwide. Contact your school, the agency, or your local state employment office for more information.

FEDERAL JUNIOR FELLOWSHIP PROGRAM

The Junior Fellowship program is for graduating high school seniors who plan to attend postsecondary school. You may work during summers or part-time during college. Agencies contact high schools for nominations. Sometimes you can gain an edge by expressing your strong interest in an environmental or natural resource Junior Fellowship to the agency and your high school counselor early in your senior year. Students qualify if they maintain a satisfactory academic record and meet certain financial need criteria.

Federal Junior Fellowship students receive excellent benefits: salary, vacation, sick leave, holidays, retirement investment, and tuition and training assistance.

COOPERATIVE EDUCATION PROGRAM

Another excellent program, cooperative education, complements periods of academic study with on-the-job employment experience with federal agencies in your field of study. Some students have developed their work projects into a master's thesis, saving time and earning a salary as they move forward in their career. Co-op students receive all the regular benefits of permanent federal employees, such as a good salary, sick leave, annual leave, holiday pay, health insurance, life insurance, and retirement investments. They may also be eligible for tuition payments, training, payment for transportation between school and work, and membership in employee credit unions and recreational centers.

This is a well-established program that agencies such as the U.S. Fish and Wildlife Service, U.S. Forest Service, National Marine Fisheries Service, Corps of Engineers, and Department of the Army have used for years to obtain and develop top talent in natural resources management.

The co-op program lets you create your own position. Co-op students generally do not "count" against an agency's limit of staff numbers. Look for an agency that needs work done and needs more help, has adequate funds, but has hit its personnel ceiling—the environmental or natural resource manager, the person with too much work and not enough help, would welcome you. Do this through an informational interview where you gather facts on the agency's needs. It is relatively easy for an agency to hire you if it has an established cooperative agreement with your school. If not, your next step is to help develop the cooperative agreement between the agency and the school. Then recruitment follows, and you are hired.

The program is open to high school, undergraduate, and graduate students who maintain a good academic record and are recommended by the school's cooperative education program. Work periods may alternate with academic semesters, or you may work part-time parallel to taking classes. After receiving your degree and ending the co-op period, you are eligible to transfer noncompetitively into a permanent, career-conditional federal position with your current employer or any other federal agency with an entry-level environmental or natural resource vacancy.

OUTSTANDING SCHOLARS PROGRAM

You may apply directly to GS-5/7 entry-level federal job vacancies if you have a 3.5 GPA or greater or graduated in the top 10 percent of your college class. This is the Outstanding Scholars program (OSP), in which your reward for good grades is a shortcut to federal jobs and exemption from some of the more bureaucratic hiring procedures.

In past years, top candidates missed tremendous career opportunities because they missed the "open window" when the U.S. Office of Personnel Management was accepting applications. Even if you made the list, competition was intense for each vacancy—some outstanding individuals had to wait two to five years or more for their first permanent job offer. Now, OSP-eligible candidates are eligible for noncompetitive appointments and can bypass the competitive lists. They have rights just like a those of permanent federal employees, who receive priority over outside candidates. Good grades really pay!

Contact the agency where you wish to work, and state that you qualify for a direct appointment under the Outstanding Scholars program. Have proof showing cumulative GPA and/or class standing. You can also search each agency's internal job vacancy announcements for opportunities.

PRESIDENT'S MANAGEMENT INTERN PROGRAM

The Presidential Management Intern program (PMI) hires about four hundred graduate students annually to two-year appointments in a federal agency. The PMI attracts talented students from a wide variety of academic disciplines who are interested in a career in developing public policy and managing programs. Some agencies hire environmental management interns. Students must be nominated by their school and be candidates for a graduate degree.

For specific information, write to

U.S. Office of Personnel Management, PMI Program, 1900 E Street, Washington, D.C. 20415.

REGISTERS

Registers are lists of persons eligible for a certain position in a certain geographic area. The U.S. Office of Personnel Management develops and manages most registers for permanent positions. OPM announces when it is accepting applications, rates each applicant, scores any required tests, and submits lists of the highest-rated candidates when requested by a hiring agency. Contact OPM regarding the various registers. The following are the current nationwide registers for entry-level positions:

BIOLOGICAL SCIENCES (fisheries, wildlife, general biology, microbiology)

ADMINISTRATIVE CAREERS WITH AMERICA (park ranger, outdoor recreation planner, environmental protection specialist, game law enforcement)

PROFESSIONAL ENGINEERING (environmental engineer)

PHYSICAL SCIENCES (environmental scientist, geologist, hydrologist, surveyor)

CASE EXAMINING BY U.S. OFFICE OF PERSONNEL MANAGEMENT

Certain federal positions do not have an up-to-date list of eligibles, and OPM accepts applications on a case-by-case basis.

DELEGATED DIRECT HIRING AUTHORITY

Some positions are unique to one or several agencies. OPM realized this and has delegated hiring authority for certain positions to individual agencies. For example, the Forest Service hires forestry technicians, and the National Park Service hires park rangers directly in many cases. Direct hire is a tremendous advantage to the agency and to temporary employees dedicated to the agency; it helps the agency elevate and promote temporary employees who were hired under other competitive announcements to permanent positions.

Contact your local Federal Job Information Center, describe the position(s) you are seeking, and ask which agencies have delegated direct hiring authority in that region. Otherwise you may miss some job openings, because OPM does not include every direct hire position in its job vacancy listings.

At the time of this writing, Congress was considering legislation that would allow, and possibly require, agencies to directly hire temporary employees to permanent positions and provide health, life insurance, and retirement benefits if they have worked for a certain number of years at the agency. This humanitarian legislation was triggered by the tragic death of a seven-year temporary National Park Service ranger who had no federal life insurance benefits to leave his young family.

PEACE CORPS ALUMNI PROGRAM

Commit to using and teaching your natural resource, ecological, engineering, forestry, or fisheries skills to people in developing countries for about twenty-seven months, as a Peace Corps representative, and you gain a world of experience. You also gain an important inside track to federal jobs—one year of noncompetitive eligibility for federal job vacancies. The international experience also provides a competitive edge as environmental jobs become more global.

EPA has already established its Office of International Activities, which provides technical assistance to developing countries. Peace Corps alumni have over two years of specialized technical assistance experience and should be prime candidates for EPA's positions. EPA's international office also works on environmental issues such as global warming, ozone depletion, hazardous waste transport, and protecting marine ecosystems, so the Peace Corps experience could lead you to a promising international career working for EPA or for other federal agencies that have international missions.

The Peace Corps experience also provides an advantage when seeking any other federal job within the United States. Contact federal agencies directly, and tell them you have noncompetitive eligibility as a Peace Corps alumnus.

WHERE TO GO FOR FEDERAL JOB INFORMATION

The U.S. Office of Personnel Management operates approximately fifty Federal Job Information Centers nationwide to serve you. Your nearest center should be listed under U.S. government in your telephone book. Centers have the latest details on competitive vacancy announcements and registers.

Career America Connection is OPM's voice mail recording of nationwide federal job vacancies and career information. It

Neil Curry, Arkansas State Parks
Earth Worker

As interpreter at Pinnacle Mountain State Park in Arkansas, Neil Curry comes in contact with at least half a million park visitors each year. "What I like most about my job is that I get to meet diverse groups of people and instill in them a sense of wonder about the natural environment," says Neil.

Now in his eleventh year working at the Arkansas state park, Neil says he has been interested in the natural environment for as long as he can remember. During his summer break in 1979, before beginning his junior year at Oklahoma State University, Neil worked as an SCA resource assistant at Glacier National Park in Montana. "That experience reconfirmed my interest in working in park or visitor interpretation. It was one of the best things I ever did," says Neil.

While completing his course of study toward a B.S. in zoology (with an emphasis on wildlife management and botany), Neil accepted a seasonal job at Chickasaw National Refuge Area in

continued

operates seven days a week, twenty-four hours a day. The only cost is for the long-distance call. Career America Connection's telephone number is (912) 757-3000.

Call the OPM's Federal Job Opportunities Bulletin Board if you have a personal computer equipped with a modem and communications software. The board has a wide variety of helpful federal career information and nationwide job listings. The nationwide bulletin board is available anytime by calling (912) 757-3100.

State employment offices also provide federal job information, but you will usually have to go into the office to review announcements and take notes—they generally do not provide copy services.

continued from page 55

Oklahoma. After graduation, Neil began working for the Arkansas State Park system as a seasonal employee, and after six months he was moved to full-time permanent status.

"People in the field told me that I would never find a job," says Neil. "The jobs are out there; you just have to go look for them. You have to be dedicated enough to volunteer to get in the door."

Throughout his professional career, Neil has been involved in various aspects of the environmental field. He has helped to maintain at least forty miles of backpacking trail and lobbied on Capitol Hill for national trails, served as president of the Audubon Society of Central Arkansas, and been recognized as Wildlife Conservationist of the Year by the Arkansas Wildlife Federation. At present, he works as a volunteer teacher for Project WILD and is a member of the National Association of Interpretation. In addition, Neil has made certain that his park is an active supporter of SCA. "We're the only park [in the state] that offers an SCA program. I realize the value of good labor," says Neil.

EXPERIENCE WITH STATE ENVIRONMENTAL AND NATURAL RESOURCE AGENCIES

Working with a state environmental or natural resources agency can provide excellent experience for federal jobs, but it is difficult to stay abreast of hiring procedures for all fifty states, which vary widely. Some state agencies recruit directly through their personnel offices or through state employment offices—sometimes both. Some states require a test; others evaluate a completed state application form.

Recruitment may be on a case-by-case basis, or agencies may open up a register to seek a pool of qualified applicants for both the current vacancy and future vacancies in the same career field. For example, the Minnesota Department of Natural Resources may recruit for a natural resources specialist II (fisheries) for the Duluth area office and use that particular vacancy announcement to recruit for other natural resource specialist II (fisheries) candidates for future vacancies anywhere in Minnesota.

Residency requirements vary. In general, agencies prefer to hire state residents. Two key factors to success in securing state jobs are networks and knowledge. Developing a good network of contacts in state agencies and having thorough knowledge of the state's ecosystems, habitats, and plant and animal species are extremely important to obtain that first state job. Agencies often require that the prospective employee become a state resident within a certain time in order to hold a state job.

Each state usually has one standard application form that applies to all permanent positions. In some cases, individual state agencies have their own forms.

TWELVE WAYS TO WIN SUMMER JOBS WITH FEDERAL AGENCIES

Winter is the time to apply for summer and/or seasonal jobs with the federal government. Most field offices of the Bureau of Land Management, Fish and Wildlife Service, and Forest Service begin summer and seasonal recruitment then. Some agencies have late December or mid-January application deadlines.

SUMMER JOBS ANNOUNCEMENT NO. 414

1. OPM publishes its Summer Jobs Announcement No. 414 in December. It provides general information on job vacancies, hiring procedures, and deadlines for agencies such as the Corps of Engineers, EPA, National Marine Fisheries Service, Agricultural Research Service, and many others.

2. Most of the Federal Job Information Centers provide a local and very important supplement to Announcement No. 414. There are approximately fifty centers nationwide, and their supplements to Announcement No. 414 list specific summer jobs in each state or region. Contact all the centers if you would take a job anywhere in the United States. Some agencies will also publish their own summer and/or seasonal recruitment announcements in addition to, or in lieu of, the OPM Announcement No. 414 Supplement. The summer positions technically are those that begin no earlier than May 12 and end no later than October 1.

3. Some seasonal positions beginning in February and lasting until November are also available at this time of year; agencies will usually open recruitment locally with a vacancy announcement.

4. In general, an SF-171 (federal application form) and OPM Form 1170/17 (list of college courses) are required for applying to these positions. Competition is extremely keen for most of them, so apply for as many as possible.

TYPES OF POSITIONS AVAILABLE

Jobs include

- ⊷ Environmental protection specialists/scientists
- ⊷ Biological technicians (wildlife, fisheries, plants)

- Forestry aides/technicians
- Landscape architects
- Archaeologists, archaeological aides/technicians
- Biologists, wildlife biologists, fisheries biologists
- Physical scientists/technicians (geology, soils, hydrology)
- Geographers (GIS)
- Soil scientists/conservationists/technicians/aides
- Recreation aides/specialists
- Laborers

As of 1993, the salary level was generally GS-2 to GS-7 ($5.95 to $10.11 per hour); assume a 2 to 5 percent increase each year. Some agencies hire GS-9 or GS-11 professionals for their regional offices or headquarters. Laborers are paid on a "wage grade" scale, which covers blue-collar workers.

OTHER AGENCY APPLICATION PROCEDURES

The Bureau of Land Management (BLM), Forest Service (USFS), Fish and Wildlife Service, and National Park Service each have their own unique hiring procedures.

5. Forest Service: National forests use the state employment service offices for summer recruitment of new employees, according to personnel specialists at the USFS Pacific Northwest Regional Office in Portland, Oregon. Local ranger districts recruit directly through local state employment offices. You should contact those offices directly for details. Some state employment offices within the same state may require different application forms. There is usually no application deadline, but apply early to increase your chances.

6. The Payette National Forest's McCall (Idaho) Smoke Jumper Base and Region 1 serial fire depot in Missoula, Montana, hire many of the USFS smoke jumpers. The McCall office, (208) 634-0700, concludes hiring for GS-5 forestry technicians (smoke jumpers) on December 31. The Missoula office, (406) 329-3511, hires forestry technicians (smoke jumpers) December 1 through January 15.

7. The BLM also recruits through local state employment offices. This agency uses the same system as the Forest Service and hires range technicians, biological technicians, survey aides and/or technicians, and others.

8. You can go to your local state employment office and tie into the Interstate Job Bank to search for some of BLM's "hard-to-fill" summer jobs nationwide. Contact local state employment offices several times per month for the more desirable positions.

The BLM Interagency Fire Center hires forestry technicians (smoke jumpers) through the Idaho Department of Employment Office, 8455 Emerald Street, Boise, Idaho 83704. The work is grueling, and the physical requirements are substantial. At least one season of wildland fire suppression experience is required.

9. Fish and Wildlife Service (U.S. FWS) hiring procedures vary from region to region. The SF-171 is the proper form to use, although there will be very few new summer hires within the Fish and Wildlife Service in the coming years.

Region 6 (Denver) says their regional office publishes vacancy announcements. Those announcements are posted at the local state employment offices and the Federal Job Information Centers. For example, the Denver FWS office has recently opened applications for thirty temporary employees at the Northern Prairie Wildlife Research Center. Most are biological technicians, GS-4/5. Other positions will become available later; application periods are brief.

10. The National Park Service (NPS) appears to be one of the few agencies that has not drastically changed summer job recruitment procedures. To apply for a park ranger position, contact the NPS directly in Washington, D.C., at (202) 343-4885 for the application package. The application deadline is January 15.

Some NPS vacancies for biological technicians, hydrological technicians, lifeguards, and other positions are advertised locally.

11. The EPA, Department of Energy, Bureau of Reclamation, Bureau of Indian Affairs, and Department of Defense generally announce their summer and seasonal positions through local job announcements. Many of their vacancies are also listed in the various OPM supplements to Announcement No. 414. The SF-171 and OPM Form 1170/17 are the standard forms. Application deadlines vary from January though April.

12. Tips for Success:

- Update and clean up your SF-171 and OPM Form 1170/17 (list of college courses). Include all courses that you will complete by the end of the school year. Make plenty of crisp copies.

- Contact a specific field office, and state that you specifically listed that office as your "most desired" location. Begin the rapport now before the office is flooded with hundreds of "look-alike" applicants.

- Use the National Wildlife Federation's *Conservation Directory* for a list of agency addresses and telephone numbers. Most libraries carry it. Call the NWF at (202) 797-6800 to purchase a copy.

- Send your applications early. Agencies often fill the best positions first.

- Check with agencies again in June. Some people just do not show up for work, and agencies must scramble to fill a few remaining positions.

THE NATIONAL PARK SERVICE

The National Park Service (NPS) employee symbolizes the promise and the problems of natural resource professionals everywhere. Getting a job in a magnificent national park like Yellowstone or Yosemite—and a ranger hat to boot—has always been the stuff of dreams. Today's Park Service, however, advises prospective employees to wake up for a reality check; consider the variety of other hats they could don in the Service; explore the variety of natural, historic, recreational, and cultural units in the National Park System—and *then* follow those dreams.

No employer holds greater promise of adventure in a natural or cultural setting than the Park Service. The National Park System encompasses 367 units and more than 80 million acres coast to coast and beyond. It includes many of America's most famous natural wonders and cultural and historic sites—from "Old Faithful" Geyser at Yellowstone to Mount Rushmore, from the battlefields of Manassas to the wetlands of the Everglades, from the dome at Yosemite to the ancient mountains in the Great Smokies, from panoramic views at the Grand Canyon to closeups at the White House.

The park system also includes many lesser known but equally fascinating areas. Included are not only national parks but also national monuments, battlefields, historic sites, lakeshores, seashores, recreation areas, and scenic trails, and units with other designations. A Park Service career could find you traveling from vast Arctic wilderness to lush forests in Hawaii, from hidden cliff dwellings at Mesa Verde to underwater trails in the Virgin Islands. You will need to adapt to more than the weather, however. Robert Cunningham, general superintendent of the Southern Arizona Group of the National Park Service, told *Earth Work*, "Regardless of whether they manage the quarter-acre Ford's Theater National Historic Site or the 13.1 million acres of Wrangell–St. Elias National Park and Preserve in Alaska, park managers and rangers need a thorough understanding of the agency's administrative policies and extensive training, in addition to a science education, to adequately preserve, manage, and interpret each park area."

The diversity of NPS units and the complexity of management issues have expanded greatly in the past three decades. Once associated with natural and cultural areas only, in the 1970s the National Park System adapted to the addition of urban national recreation areas at Golden Gate in San Francisco, Gateway in New York, and Cuyahoga Valley in Ohio. Then the Alaska National Interest Lands Conservation Act of 1980 more than doubled its acreage to incorporate our nation's only remaining vast wild areas and the wildlife they shelter. About two-thirds (more than 53 million acres) of the acreage under NPS management today is found in Alaska, yet the parks there are markedly understaffed.

Along with the U.S. Fish and Wildlife Service and the Bureau of Land Management—other agencies profiled in this book—the National Park Service is a bureau of the federal Department of the Interior. The NPS has approximately fourteen thousand full-time permanent employees and six thousand seasonal or part-time employees—a relatively small fraction of the estimated hundred thousand park and recreation employees in state, municipal, and other park sites nationwide. Many of these other employees enjoy good working conditions and better pay.

NPS jobs, however, still seem more glamorous to many job seekers—glamorous enough for coverage in the pages of *National Geographic*. On the occasion of the NPS Diamond Jubilee in August 1991, for example, *National Geographic* assigned one of its senior staff writers to prepare extensive portraits of NPS employees. The accompanying photos ranged from a shot of Mike Danisiewicz, a ranger at Saguaro National Monument, and his wife, Patte, a biological technician, leaning against a twenty-five-foot cactus; to one of Ailema Benally at Hubbell Trading Post National Historic Site, surrounded by a room full of rugs woven by Navajo women like her grandmother; to one of biologist Mary Meagher observing bison in Yellowstone. In describing the latter, *National Geographic* underscored that "that's not just a bison in the spotting scope research biologist Mary Meagher holds, it's a last vestige of frontier."

The publicity and the public affection for parks reflect a symbolism that goes beyond colorful people and colorful places to explain why NPS jobs are enviable: More than any other agency, the Park Service symbolizes America's wild heritage. For every NPS employee who makes it into print, there are several thousand others. Every day they quietly contribute to the protection and understanding of our nation's living and historic treasures—whether by making new discoveries or interpreting the plants, wildlife, and culture of the parks for the public. Several years ago, for example, park ranger Brent Wauer made botanical history by discovering a new species of violet in Guadalupe Mountains National Park in Texas. And Karen Wade has made personnel history as the first woman superintendent of our largest national park, Wrangell–St. Elias.

Earth Work has routinely profiled career professionals in the agency—including unsung heroes and heroines—covering the problems and career issues that do not often make it into the general press or the coffee table books but that nonetheless concern NPS employees.

Park employees are plagued by relatively low pay and live and work in parks where the buildings and roads are deteriorating. Robert Cahn, author of a Pulitzer-prize-winning series on the national parks that appeared in the *Christian Science Monitor*, reported, "Most parks are falling badly behind in

Useful Publications

The National Recreation and Parks Association publishes a number of newsletters, magazines, and special reports of use to national and state park professionals as well as recreation professionals. A pamphlet, "Natural Resource and Park Management," provides useful information on job titles and descriptions, education and training, professional certification, hours and earnings, and the employment outlook. *Park and Recreation Job Opportunities Job Bulletin* is published on the first and fifteenth of each month and lists largely state and local park and recreation jobs ($5 per copy). NRPA also publishes a survey of salaries and benefits for state park and recreation personnel in fifty states and a calendar of professional events of interest to current and prospective national and state park employees. For more information, write

National Recreation and Parks Association, 2775 S. Quincy Street, Suite 300, Arlington, VA 22206-2204

essential maintenance, their natural and cultural resources are being threatened, and many are having to lay off seasonal rangers or curtail visitor services and interpretation due to lack of sufficient operating funds." Staff and funding have not begun to keep pace with a dramatic growth in park visitors and acreage over the past two decades.

Rangers in Grand Canyon National Park, for example, live in thirty-year-old trailers. NPS employees also must deal with problems never foreseen by the Park Service's founders—such as crime, crowding, and threats from outside park boundaries. Their increasing responsibilities in those areas subtract from the time rangers have to provide interpretation and guidance to visitors and from the ability of the NPS to carry out its scientific work. George Berklacy, the respected former public affairs director of the agency, has said, "I've been with the Park Service thirty years, and I miss the way it used to be, when a ranger would come by your campfire, have a cup of coffee with you, and just talk about the scenery."

Times have changed. The summer of 1993 brought the first-ever shooting of a ranger in Yosemite National Park. It also brought a new, outspoken National Park Service director, Roger Kennedy, who decided to wear the ranger's uniform himself as a sign of his promise to increase the nation's pride in and support for our national parks. He says the Park Service will become more activist in behalf of both the parks and park employees: "Hunkering down is no longer the order of the day," Kennedy explained in an October 27, 1993, interview in the *Kansas City Star.* His top priorities: "We've got people, that's the Service. And place, that's the system. . . . How do we assure people that if they give their lives to this Service they will be treated respectfully, paid decently, given opportunities to grow, fulfill themselves? . . . The system has become shabby, tattered, overused, underprotected, despite what is really dedicated service by the Service. We need to do everything we can to protect that heritage." Asked what he plans to do about the issue of low salaries, Kennedy told the *Star,* "There are the devices of conventional leadership, which are inexpensive and have worked since Peter the Hermit or George Washington. Then there are things that have to follow closely, which are real, such as the reclassification of salaries, which are disgracefully behind what they should be for commensurate work."

Kennedy's historical perspective and sense of drama have been well honed during his career. He served as director of the Smithsonian Institution's National Museum of American History from 1979 to 1992. Before putting on his ranger hat, he wore the hats of a vice president of the Ford Foundation, a White House correspondent and producer for NBC, a lawyer and special assistant to the secretaries of labor and education, and a bank executive. He has written seven books, including *Rediscovering America,* the basis of his television series, *Roger Kennedy's Rediscovering America.* The *Washingtonian* magazine recently named him one of the twenty-five most intelligent people in Washington, D.C.

As one of his first steps in office, Kennedy promised to carry out and intensify an agenda for change devised by Park Service employees and other park leaders in follow-up to a study of park needs begun during the Seventy-fifth Anniversary Year of the NPS in 1991.

JOBS IN THE PARK SERVICE

The Park Ranger Series

The park ranger series is the largest single occupational category in the NPS, accounting for roughly one-quarter of the agency's workforce. The jobs in this series are as varied as the national park units themselves. The protection and use of historical, cultural, and natural resources in these units involves familiarity with the special conservation and management needs of wildlife, forests, seashores, varied recreation units, historical buildings, battlefields, and archaeological properties—to name a few. Increasingly, it also involves law enforcement activities.

Since 1985, the series has incorporated "park ranger" and "park technician." It lumps together many activities of different skill levels; the professional duties covered break down into the major activities of park conservation, public safety, public use and recreation, interpretation, public relations, law enforcement, and park management.

The park ranger series is not recognized by the Office of Personnel Management (OPM) as a professional series, and NPS rangers often make less than their counterparts in other agencies. OPM does have a written description of the experience and education required to qualify for an entry-level park ranger position—called a qualification standard. This standard, however, has changed a number of times. Four decades ago, entering rangers were required to have a bachelor's degree with "the major study in forestry, conservation, physical geography, or wildlife management, or in the natural history or field phases of biology, geology." By 1990, the standard had changed so that the bachelor's degree requires only "twenty-four semester hours of related course work in the fields of natural resource management, natural sciences, earth science, police science, social sciences, museum sciences, business administration, public administration, behavioral sciences, sociology, or other closely related subjects." Robert Clay Cunningham, general superintendent of the Southern Arizona Group of the NPS, reports that expansion of the acceptable courses of study and the lack of a required major field have resulted in a workforce of park rangers only half of whom hold degrees in natural or cultural resources studies. Moreover, many have substituted experience for degrees.

Currently the NPS is examining ways to improve pay and working conditions for its employees and reevaluating the qualification standard. Reclassification of the park ranger position is being studied intensively. In addition, the Clinton administration is targeting personnel as an area of reform government wide. The gist of the change is that federal agencies will be granted authority to design their own personnel programs.

Given the possible changes ahead, those who are still in school and are potentially interested in a park ranger career would be wise to choose one of the majors based in natural resources or cultural resources or concentrate in park and recreation management.

Because of the popularity of park ranger jobs and the intense competition for relatively few openings, a more detailed description, "Becoming a Park Ranger," appears at the end of this section on the NPS. Although you may picture yourself in Yellowstone or Yosemite, bear in mind that more than half of the nation's park rangers work east of the Mississippi, and although they often work outdoors, they also work in offices, especially as they advance to district ranger, park manager, and other managerial positions. However, if you still have your heart set on being a ranger, don't give up—these tips will give you a head start.

Other Series

Although rangers are the most familiar, the NPS workforce is quite varied. Jobs available in the Service include positions in the U.S. Park Police, administrative careers, maintenance and trade positions, land acquisition and concessions specialists, and some positions in the biological sciences. Persons with backgrounds in archaeology and history and, to a lesser degree, sociology, geography, and anthropology, conduct cultural resource programs. Museum jobs, though few in number, are particularly exciting because of the varied collections of the NPS.

For a twenty-page booklet on NPS careers, "U.S. Department of the Interior National Park Service Careers," write

National Park Service, P.O. Box 37127, Washington, DC 20013-7127.

Also contact National Park Service Regional Offices:

Mid-Atlantic Regional Office, National Park Service, 143 S. Third Street, Philadelphia, PA 19106, (215) 597-7013

North Atlantic Regional Office, National Park Service, 15 State Street, Boston, MA 02109, (617) 223-5001

Southeast Regional Office, National Park Service, 75 Spring Street, SW, Atlanta, GA 30303, (404) 331-5185

Midwest Regional Office, National Park Service, 1709 Jackson Street, Omaha, NB 68102, (402) 221-3431

Southwest Regional Office, National Park Service, Old Santa Fe Trail, P.O. Box 728, Santa Fe, NM 87501, (505) 988-6388

Rocky Mountain Regional Office, National Park Service, P.O. Box 25287, Denver, CO 80225, (303) 969-2500

Western Regional Office, National Park Service, 600 Harrison Street, Suite 600, San Francisco, CA 94107, (415) 744-3876

Pacific Northwest Regional Office, National Park Service, 83 South King Street, Suite 212, Seattle, WA 98104

National Capital Region, National Park Service, 1100 Ohio Drive SW, Washington, D.C. 20242, (202) 619-7005

Alaska Office, National Park Service, 2525 Gambell Street, Anchorage, AK 99503-2892, (907) 257-2690

THE VAIL AGENDA

In 1991 the National Park Service celebrated its seventy-fifth anniversary by dedicating a new museum about park rangers of yesteryear at Yellowstone and beginning an extensive process of soul-searching about how to improve the future for today's rangers and other NPS employees. In an October 1991 meeting in Vail, Colorado, a number of leaders convened to examine input from NPS employees nationwide and come up with an action plan. As one of his first steps, the new NPS director, Roger Kennedy, promised to carry out the Vail Agenda vigorously.

At a November 1993 "ranger rendezvous" in Virginia Beach, NPS Deputy Director John Reynolds pointed out that "the Vail Agenda meshes very nicely with the administration's efforts to reinvent government. It is our good fortune that at the same time we are trying to establish new programs, the reinventing government initiative is giving us the flexibility to experiment with new ideas and make responsible, 'real-world' changes."

The Vail Agenda aims to bring the Park Service back to its original purpose as stated in the 1916 law that created the agency:

to conserve the scenery and the natural and historic objects and the wildlife therein and to provide for the enjoyment of the same in such manner and by such means as will leave them unimpaired for the enjoyment of future generations.

Published recently, the Vail Agenda calls for accomplishing that purpose through the following strategic objectives:

RESOURCE STEWARDSHIP AND PROTECTION. The primary responsibility of the National Park Service must be protection of park resources.

ACCESS AND ENJOYMENT. Each park unit should be managed to provide the nation's diverse public with access to and recreational and educational enjoyment of the lessons contained in that unit, while maintaining unimpaired those unique attributes that are its contribution to the National Park System.

EDUCATION AND INTERPRETATION. It should be the responsibility of the National Park Service to interpret and convey each park unit's and the park system's contributions to the nation's values, character, and experience.

PROACTIVE LEADERSHIP. The National Park Service must be a leader in local, national, and international park affairs, actively pursuing the mission of the National Park System and assisting others in managing their park resources and values.

SCIENCE AND RESOURCES. The National Park Service must engage in a sustained and integrated program of natural, cultural, and social science resource management and research aimed at acquiring and using the information needed to manage and protect park resources.

PROFESSIONALISM. The National Park Service must create and maintain a highly professional organization and workforce.

BECOMING A PARK RANGER

Few jobs elicit the envy of the working public like that of park ranger. Jonathan Jarvis, superintendent of Craters of the Moon National Monument in Idaho, compares the image to the reality: "An American icon, the park ranger evokes a romantic image of the mountain-wise hero, ready to save a fawn from a raging forest fire, to rappel to the rescue of a trapped climber, and to offer boundless knowledge of the out-of-doors to the park visitor. In that enviable job, the park ranger must constantly answer the second most commonly asked question of the profession: How do you get to be a ranger? (the most frequent question is, of course, Where are the restrooms?)."

Some believe that all it takes is a good education, a few seasons working as a park ranger, and—bingo!—you have your own horse and a hundred thousand acres of wilderness to patrol. Well, it just does not work that way. Here, we will try to separate myth from reality and explain the somewhat daunting process of becoming a permanent park ranger with the National Park Service.

The NPS is an equal opportunity employer and makes it difficult to become a park ranger regardless of gender or ethnic origin. And there is a clear distinction between the park ranger as federal employee of the NPS, part of the Department of Interior, and the forest ranger as a federal employee of the Forest Service, a division of the Department of Agriculture. The primary difference according to Jarvis is that "park rangers work in national parks, forest rangers work in national forests, and park rangers are usually better-looking than forest rangers."

What is a park ranger? The title is a specific classification of the Office of Personnel Management (OPM), in the series GS (General Schedule) 025. Grades span from GS-025-03 to GS-025-16. In dollars that equals a rough salary range of $14,000 to $73,000, depending on locality as well as grade. Duties range from collecting entrance fees at Shenandoah National Park in Virginia or leading tours of the Frederick Douglass Memorial Home in Washington, D.C., to driving patrol cars after speeders on Blue Ridge Parkway in North Carolina, conducting wildlife studies in Yellowstone National Park in Wyoming, or serving as a superintendent at North Cascades National Park in Washington. Also, many nonfield positions in the ten regional offices of the NPS are also classified as park rangers, so obviously not all park rangers ride a horse, give campfire talks, or carry a gun.

Current OPM standards for park rangers are classified under the Administrative Series, and therefore do not require a college degree. These standards are a result of the wide variety of work carried out by park rangers and are now being reviewed for probable change. Out of the fourteen thousand permanent positions at the NPS, fewer than a third are classified as rangers. Other positions include mechanics, typists, landscape architects, biologists, archaeologists, historians, curators, contract specialists, and budget technicians.

Seasonal vs. Permanent: The Difference

At the NPS, there are seasonal rangers, and there are permanent rangers—the differences are distinct. Permanent rangers are hired through OPM or under one of the special hiring authorities (see "Federal Hiring Procedures" earlier in this chapter). Such rangers receive insurance, health, and retirement benefits and the appointments have no specified end. Permanent positions are classified into categories "career conditional" and "career status." When an employee has been hired into a permanent position but has not worked a full three years, he or she is often referred to as career conditional. After three years of satisfactory performance, an employee is granted career status with the federal government.

Conversely, seasonal park rangers are not hired through OPM, receive no benefits, and have appointments with a specific time limit.

There are also career-seasonal positions that are permanent but subject to furlough. These positions receive all the benefits of permanent status but may include up to a maximum of six months of nonpay status. Many entry-level ranger jobs fall under this category.

The NPS hires approximately four thousand seasonal park rangers per year for temporary appointments. They are hired from the Seasonal Register, a list of applicants used by all NPS areas to fill temporary positions. Applications must be sent to Washington, D.C., before January 15; each year you can apply to only two parks. Seasonal employees develop no career status from their work time, and grades are usually GS-4 to GS-5 (with a rare GS-6 and GS-7). Most seasonal positions are for the summer months only; a few parks hire for the winter. Seasonal rangers have no guarantee of returning the following year. Though these jobs are rewarding and build valuable experience, they do not provide a direct avenue to becoming a permanent park ranger.

Most permanent park ranger positions are advertised through an NPS vacancy announcement and are open only to those with career or career conditional status. In other words, you have to be a permanent employee within the federal government or fall under one of the special hiring authorities (in the following list) to be hired as a permanent park ranger. A seasonal employee, no matter how well qualified, cannot be hired from a vacancy announcement for a permanent park ranger position.

Becoming Permanent: The Hiring System

OPM is the federal agency responsible for administering vacancy announcements and written exams for all government offices that fill positions from outside the federal workforce. You should not wait until you hear about a job opportunity but rather get on the OPM register in advance.

OPM issues a Federal Job Information listing every two weeks. Included are local, regional, and national vacancy announcements with indications of which positions require a written exam. Once you have submitted application forms and have taken the test, you will be given a numerical rating based on score, experience, training, and education. Eligible applicants are placed on "lists of eligibles" or "registers" in order of rating. Eligible nondisabled veteran applicants get an extra five points; disabled veterans get an extra ten points. The NPS, if it wishes to consider candidates from outside the federal government, requests a list of eligible candidates (certificates) from OPM. It then may select any one of the top three available candidates on the certificate; however, a veteran cannot be passed over without approval from OPM.

Applicants for all park ranger positions—regardless of whether they include law enforcement—must take the Law Enforcement and Investigation examination of the OPM Administrative Careers with America program. To apply, contact your local OPM area office (listed in the government pages of your telephone book) and arrange to take the "Law Enforcement and Investigation Examination Cont. NO. MC0032." Note: It may take up to three months to schedule the exam and another several months to receive your rating. Once listed, you remain on the register for one year and can be extended by request for a second year.

Special Hiring Authorities

Getting on the federal job register is not the only way to qualify for a ranger position. Here are other ways to get in.

Cooperative education program. This approach can be rewarding for both the individual and the park. It is a formal agreement between a park and college or university for specific positions. The school announces the openings, and the park makes its selection from the list of applicants. Students spend at least two summers and often another quarter working at the park, employed as a GS-4 or GS-5, completing a specific park project. The program can be structured either as an undergraduate or graduate (usually part of a master's) program.

Upon successful completion, the student is given ninety days of career status with the federal government and can apply or compete for any permanent job for which he or she is qualified. This program works well with a "mentor" (see section on mentors below) who can hold open a vacancy until the student completes the co-op program. If you are an undergraduate and/or contemplating a graduate degree, contact the park of your choice for details.

Veteran Readjustment Appointments (VRA). Veterans can be appointed to permanent positions, up through GS-9 or Wage Grade-9, without OPM approval. To be eligible, a veteran must have no more than fourteen years of schooling.

Selective placement. Agencies such as the NPS have affirmative action programs for handicapped and disabled veterans, giving them authority to hire severely handicapped persons without OPM approval.

Steve Sarles, National Park Service
Earth Worker

Steve Sarles traveled from coast to coast for several years working as a seasonal employee with the National Park Service before finally obtaining permanent status. But today he is supervisory ranger for Yellowstone National Park. "A person looking for work with the National Park Service must be open to all kinds of opportunities," says Steve. "If you restrict yourself to a certain park, the chances are, you are not going to get into the Park Service." He adds, "Everyone wants the big parks, so you've got to be flexible." He also suggests that the Park Service is more competitive "because of the many qualifications a park ranger must have these days—law enforcement, emergency medical services, interpretation," and more.

Steve's career exemplifies his sentiments. Upon receiving his bachelor's in environmental sciences in 1977 from Davis and Elkins College in West Virginia, Steve was uncertain about what he wanted to do with his degree. He knew he wanted to work outside and contribute to helping the environment. On the advice of a

continued

Peace Corps, Vista, and Action volunteers. After serving a full term of satisfactory service, usually two years for Peace Corps and one year for VISTA or ACTION, you can apply for a position without going through OPM.

Minimizing the Competition

To successfully land a permanent park ranger job takes more than just completing an application with OPM. The competition is frequently intense even for jobs with the highest qualifications; therefore, minimizing the competition is essential. Here are insiders' tips on getting an edge.

First, consider "going urban." With good qualifications, it is often easier to land a permanent park ranger job in an urban park than it is in a classic rural park, because the lower pay scale in rural parks leads to higher turnover. However, if your ultimate goal is to work in a large rural park, then you had best obtain seasonal experience in one before getting a permanent job at an urban area. It is a poor idea to "go urban" for the ninety days needed to gain career conditional status and then abandon the commitment. It is unfair to the urban parks and may burn bridges with the NPS personnel.

Another route is to obtain career status in another somewhat related federal agency, such as the Army Corps of Engineers, the office of the U.S. Geological Survey (USGS), or the Bureau of Land Management. Traditionally, the NPS has a bias toward hiring its own; therefore, applications from another agency's personnel may get the last look. The way to stand out is to have experience with the NPS as a seasonal employee in addition to the permanent employee experience in some other agency.

Other agencies have important jobs and missions too, so give them an appropriate amount of time before you jump ship. An early departure can impair other agency programs.

The Mentor: An Essential Ingredient

Working seasonally year after year at Yellowstone or Yosemite until you just "fall into" a permanent job has great appeal. It does happen—but more often than not, it only results in years of seasonal employment, low pay, poor living conditions, and eventual frustration.

The key is having a mentor who understands the NPS and OPM personnel system and has the wherewithal to make it

continued from page 68

professor, Steve applied for a resource assistant position with SCA, and the following spring was placed in Grand Canyon National Park for his first NPS position as an interpreter.

After his SCA experience, Steve traveled across the United States working seasonally with the Park Service in numerous interpretative positions—wherever there was work. He even spent a season working as an interpreter for the U.S. Forest Service and spent six weeks at the Law Enforcement Academy. Then in 1981, after a few weeks' work as a seasonal ranger at Yellowstone National Park, Steve was finally offered a permanent position as an interpreter at Prince William National Forest in Virginia. A year later, he moved back to Wyoming, accepting permanent work at Yellowstone, where he has been ever since.

His current duties are many, including overseeing the SCA program in the park, managing the Youth Conservation Corps, serving as medical services coordinator, and supervising the Visitor Center.

Steve credits his career path with the Park Service to his experience in the Grand Canyon. "My SCA experience

continued

work to your advantage. A mentor can request an OPM register for an existing park ranger vacancy, sponsor a co-op position, utilize the special hiring authorities, or hire you into a clerk-typist job and later convert the position to park ranger. Mentor relationships usu-

continued from page 69

was the single most influential event in determining my career direction with the Park Service. It was a great experience. I learned all about national parks and the Park Service. And I met lots of great people."

ally are developed when your performance as a seasonal employee stands out from the crowd and makes that supervisor want to go out of his or her way to bring you into the NPS.

Summary

To become a park ranger requires education, field experience, patience, knowledge of the system, and determination. With this combination, you will succeed. Take heart that in spite of the larger-than-life mystique, park rangers are people, and they do eventually retire, creating opportunities for the up-and-comer. The NPS has an important and noble mission, and needs the brightest and the best.

U.S. FOREST SERVICE

In 1991 the Forest Service celebrated its centennial and entered its second century in the throes of either a renaissance or an identity crisis, depending on the commentator. Acknowledged as the most professionally trained land management agency, the Forest Service is currently helping its managers deal with problems that were not typically a central part of their college curricula. The issues run the gamut, from timber harvests and grazing permits to preservation of wilderness and improvement of watersheds. Environmentalists charge that excessive logging has turned many national forests into tree farms, to the detriment of wildlife, while the timber industry warns of a timber crisis unless more trees are sold from public lands.

The National Forest System includes 156 national forests, 20 national grasslands, and 9 land utilization projects located in forty-four states, Puerto Rico, and the Virgin Islands. Lands managed by the Forest Service encompass such diverse areas as mountain glacier fields, forests, range and grasslands, lakes and streams, and tropical rain forests. Altogether, these lands make up 8.5 percent of U.S. land area, but provide habitat for half of all the nation's big game. National forests are more important than ever to Americans. With 50 percent of the nation's softwoods, 40 percent of its recreation lands, 50 percent of its big game, and 50 percent of its cold-water fisheries, the national forests are precious resources with people fighting over them, some of whom have devised grand strategies to get more than their fair share.

Unlike national parks, which are managed primarily for land and wildlife protection and recreation, national forests are intended for multiple-use management. Within a single forest, the uses can include water, forage, logging, wilderness, wildlife habitat, recreation, and mining. The Forest Service prepares integrated land and resource management plans for each national forest and must weigh conflicting public opinions about different uses of a forest against the agency's mandate.

As set forth in law, the Service's mission is to achieve good land management under the sustainable multiple-use management concept to meet the diverse needs of people. The Forest Service advocates a conservation ethic in promoting the health, productivity, diversity, and beauty of forests and associated lands. In addition, the mission of the Service includes obligations to protect and manage the national forests and grasslands and to provide technical and financial assistance to state and private forest landowners, as well as to cities and communities seeking to improve their natural environments. In the process, the Forest Service strives to achieve the human resources goal of its mission by providing work, training, and education to the unemployed, underemployed, elderly, youth, and disadvantaged. Finally, the agency provides international technical assistance and scientific exchanges to sustain and enhance global resources and encourage good land management. Jack Ward Thomas, President Clinton's newly appointed Forest Service director, believes that this mission will lead to what the Service should be doing "in the next century . . . reestablishing itself as the conservation leader of the world."

The Forest Service has undergone significant changes over the last hundred years. As recently as 1959, for instance, *The Forest Ranger,* by Herbert Kaufman, described the Forest Service as a forestry and engineering organization. The major organizational debates for at least a century were between foresters and engineers. The book describes rangers as militaristic and loyal. Kaufman writes that the Service indoctrinated its employees so thoroughly that if a problem was on the table, staff would all come up with the same answer.

According to former Forest Service director Dale Robertson, however, today's Service is considerably different: "Today's Forest Service has retained a strong culture; we look at ourselves as a family. There's a lot of caring about other people. And now our family is diversified. Sociologists, economists, fishery biologists, botanists, and archaeologists are rapidly joining the more traditional engineers and foresters. And it's no longer a white man's world." The Forest Service is "really working hard to

Mary Ann Freeman Chambers, U.S. Forest Service
Earth Worker

Coming from Bayonne, New Jersey, Mary Ann Chambers did not know what to expect when her SCA supervisors told her in 1978 that she would be working as a resource assistant eighty miles from the nearest town: "I couldn't even imagine what that would be like, but I loved it, and it started me off on the right foot."

She served as an SCA volunteer for the Canyonlands National Park, Needle District, in Utah, doing interpretive programs and park patrols. Mary Ann decided to move out west as a result, getting work as an NPS seasonal at Arches National Park and transferring to Utah State University, where she received a B.S. in forest recreation in 1983. Under a cooperative education program, she worked for six months at a time in Shasta National Forest in California and studied for six months in Utah.

Mary Ann earlier worked at Routt National Forest for four years and now is district planner for the Boulder District of Arapaho and Roosevelt National Forests in Colorado. She

continued

attract minorities and women." According to Robertson, "No natural resources organization in the world" can match the Forest Service's "rich diversity in disciplines, the number of women, and added minority diversity."

continued from page 71

writes environmental analyses and environmental impact statements and supervises the interpretive program. "Public involvement is the best part of it," she says. "I enjoy explaining what we're doing."

The new Forest Service calls for a new kind of employee, one who goes beyond the image of a lonely ranger escaping the rat race in the outdoors, peacefully contemplating the woods. Robertson says, "Unless you're pretty tough and can take criticism, don't get into a managerial position in the Forest Service. There are still jobs where you don't have to deal with conflict, but line officers and professional staff do. A Forest Service employee must know the technical stuff and have a feel for the land, a feel for the resources."

The Forest Service employs over twenty thousand professionals, including over six hundred rangers and more than five thousand foresters. The "technical stuff" former chief Robertson referred to is the expertise of the wide array of professions and degrees that are compatible with working for the Forest Service. Professionals include archaeologists and anthropologists; biologists; ecologists; botanists; range conservationists; engineers; landscape architects; architects; civil engineers; geologists; wildlife refuge managers; and fish and wildlife administrators. General and wildlife biologists are the third most prevalent professionals in the Forest Service.

When President Clinton named Jack Ward Thomas his new Forest Service chief, the move was widely hailed as marking a new era in which the Forest Service would put more distance between itself and the timber industry over issues such as ancient forest wilderness.

For information, contact

U.S. Department of Agriculture Forest Service, Washington Office, 14th and Independence SW, P.O. Box 96090, Washington, DC 20090-6090

U.S. FISH AND WILDLIFE SERVICE

The U.S. Fish and Wildlife Service (FWS) has never had the same degree of public recognition as the National Park Service and other government agencies dealing with the environment. However, those who do take the time to find out about FWS and are eventually hired often find a rewarding opportunity to work directly toward preserving and restoring America's wildlife.

FWS is the federal agency charged with principal responsibility for conserving and enhancing America's fish and wildlife. The agency traces its beginnings to 1871 when Congress established the U.S. Fish Commission to study declines of food fishes and recommend methods for reversing the decrease.

Approximately seven thousand people currently work for FWS at locations including the headquarters in Washington, D.C., eight regional offices, and more than seven hundred field units. These units include national wildlife refuges and fish hatcheries, research laboratories, and field offices. Wildlife biologists, research biologists, fishery biologists, and refuge managers fill the majority of positions. Additionally, there are limited opportunities for general biologists, botanists, foresters, and ecologists.

FWS provides advice to other federal agencies, states, industry, and the public about conserving habitat potentially threatened by development. Staff biologists assess the potential impacts of projects requiring federal funding or permits, such as dredging activities or oil leasing. They also recommend ways to avert, minimize, or compensate for negative impacts on fish and wildlife.

FWS also leads federal efforts to protect endangered species. Working with scientists from federal and state agencies, universities, and other organizations, agency biologists develop "recovery plans" identifying actions needed to save listed species.

Candidates attracted to FWS differ from those attracted to the National Park Service, according to Martha Higgins, assistant to the personnel officer at the national headquarters. "Most of our candidates are geared more toward actual refuge management and wildlife management, fisheries, and hatchery activities," says Higgins. "I think most of the people who go to the Park Service are interested in positions as rangers, outdoor recreation planners, or landscape designers."

Openings for FWS positions are few, and competition is tough. Most employees come through the competitive hiring process of the Office of Personnel Management (OPM). However, this is not the only avenue available to individuals seeking a career with FWS. One option is the Cooperative Education Program, a partnership with universities throughout the country. Interested students must apply for admittance into the program. Once accepted, students work up to twenty-six weeks for FWS during their school career, including at least one complete semester. After completing the

New FWS Director Emphasizes Ecosystem Management

In 1993 President Clinton appointed Mollie Beattie director of the U.S. Fish and Wildlife Service. During her inaugural meeting with employees, Beattie stated that one of her goals is to increase the diversification of the workforce by hiring and promoting women and minorities. She emphasized her objective of making the agency the international leader in ecosystem management and conservation of biodiversity. The new director said the Service will seek to make the national wildlife refuges—which had increased to 491 in number by December 1993—"anchor points for biodiversity and ecosystem management." She also noted that "people need to understand that the choice between people and animals, between the economy and endangered species, is a false one."

program and their degrees, individuals have 120 days to find a job noncompetitively with FWS without going through OPM.

Those successfully completing the program with good reviews can expect to find a job with the Service, according to John Schroer, manager of Chincoteague National Wildlife Refuge in Virginia. "People who do good jobs in co-ops will usually be placed in positions as soon as they get their degrees," he says.

"One good thing about the co-op program is that if you've done well, the agency will hire you full time," agrees Dr. Mamie Parker, chief of division of federal activities for FWS Region III. "Whereas if you walked in off the street, you'd have to get on the Federal Register and be picked up. For people outside of the federal government, it's really difficult to compete for jobs. Once our agency invests that time and money in you, I think we see it as a commitment to try and keep you. It's almost like a family support system, and that's pretty valuable. I don't see people bragging about that in a lot of other organizations," Parker adds.

A smaller number enter FWS through temporary hiring. Ilene Grossman, for example, received her appointment "NTE," not to exceed a year. Many temporary employees stay without having to go through the competitive hiring process. "What they often do is simply renew you as a temp or else make you permanent, but it certainly eases the hiring process if they hire you as a temporary and then make you permanent later on," Grossman explains.

During John Turner's tenure as the agency's director, FWS launched new programs aimed at recruiting women and minorities. Women currently account for 38 percent of the agency workforce, minorities about 10 percent. FWS Chief for Human Resources Jerome Butler said that these figures, when compared to other agencies, were "at least equal or higher when it comes to women and probably a little lower . . . when it comes to minorities." He also said that women and minorities were more underrepresented in higher-level positions, with higher percentages employed in lower-level and administrative jobs.

Butler explained that part of the reason for the low number of minority employees is that none of the historically black colleges offers programs in wildlife biology. FWS, he explained, was working to change this by "working with these colleges to set up programs . . . to bring minorities into wildlife manage-

Jane Nicholich, U.S. Fish and Wildlife Service
Earth Worker

Jane Nicholich definitely finds her work with the U.S. Fish and Wildlife Service rewarding. "Ensuring the survival of endangered species is personally satisfying," emphasizes the 28-year-old from New Jersey. Jane, who has always loved birds, particularly likes her current position as crane flock manager at Patuxent Wildlife Research Center in Maryland. "I enjoy my work because I get to work so closely with the birds," she says.

Patuxent Wildlife Research Center is one of only thirteen national research stations in the United States operated by FWS. Patuxent conducts research on contaminated, migratory, and endangered species. As crane flock manager, Jane is involved in the captive breeding program, which tests different rearing and release techniques for the endangered whooping crane and the sandhill crane. She is directly responsible for raising the birds for research and assisting the research biologists.

Jane, a 1986 graduate of Allegheny College in Pennsylvania with a B.S. in general biology, was both an SCA

continued

ment and fish and wildlife biology careers." He said that the agency had developed a really aggressive program under Director Turner.

The agency has been much more successful at recruiting women for managerial positions. "Our biggest success has been with women. We have three new refuge managers who are women. More important, we've got some women who are deputy regional directors and others who are [regional] heads of wildlife with responsibility for all the refuge systems in the region," Butler said.

One program at FWS, the Career Awareness Institute, provides an opportunity for women and minorities to take (and gain college credit for) courses they could not get at their own universities. In most cases they do not have such a program at their school. In the event that they do, the program allows them to see the opportunities in biology. Students complete a course called Introduction to Fish and Wildlife Biology, which is offered summers at Tennessee Tech University and is worth three hours of undergraduate credit. The first half is spent in the classroom. The second half is spent working (for pay) in the field for the FWS, at a refuge, a regional office, or laboratory.

continued from page 74

resource assistant and a High School Program participant. After her college graduation, Jane traveled to Salem, Oregon, as a resource assistant to work with the Bureau of Land Management on the spotted owl issue. She then took a volunteer biologist position at Long Point Bird Observatory in Southern Ontario monitoring songbird migration. In 1988, she joined the staff at Patuxent.

As a former volunteer herself, Jane believes volunteering is a good way to get your foot in the door and find out what is out there. "Try different experiences in different areas of the country and different types of work," suggests Jane.

Flexibility is a key requirement of those applying for FWS employment. Those who are hired can expect to be relocated numerous times during their careers. John Schroer is at his ninth refuge, his fourth position as manager. "I think the norm is five to ten refuges," he says. He feels this mobility has been beneficial to him as a manager. "It's important to be mobile in that you gain the experience from making the move and looking at different situations. If you move around a lot early in your career and see a lot of different things . . . I think you're a better-rounded manager."

Probably the most well known section of FWS is the National Wildlife Refuge system, the world's largest collection of lands reserved specifically for wildlife. The 460 refuges contain more than 90 million acres of habitat. More than half the total is in Alaska; the rest is spread across forty-nine states and several territories. This is more land than is contained in the National Park System. Employment at a refuge offers a real "hands-on" opportunity to work with wildlife and habitat.

Some refuge personnel come from other agency positions; some are new hires. "We get both, depending on the position," explains Schroer, "With higher level jobs like primary assistant manager, or chief outdoor recreation planner, or head biologist—you would normally find those people who have been stationed at other refuges before coming to Chincoteague. But we've also got the trainee-level positions and the lower-graded outdoor recreation planners and biotechs. A lot of the time, those are new employees."

Chincoteague, like most refuges, finds new employees through both the Office of Personnel Management and cooperative education programs. "It depends on the position," Schroer explains. "A lot of them come through the competitive program. Those in the professional series, such as refuge manager, wildlife biologist, and outdoor recreation planner, can come through the cooperative education program where we hire college students. Right now we have three college students who are on co-op at Chincoteague."

Those people who take the time to find out about FWS and work there often make the agency their career. "We have some very senior people," said FWS human resources chief Butler. "People love what they're doing because of the resources. It's unbelievable—they're the most dedicated people I've ever met. I think some of them would work for nothing because they love what they're doing."

For more information, write for "Careers with the U.S. Fish and Wildlife Service," a beautifully illustrated twenty-page booklet published by FWS:

Public Inquiries, U.S. Fish and Wildlife Service, Room 130 ARLSQ, 1849 C Street NW, Washington, DC 20240.

U.S. BUREAU OF LAND MANAGEMENT

Once called "the lands no one wanted," the millions of acres managed by the Bureau of Land Management are fast becoming the lands everyone wants. No longer the little-known domain of only miners and ranchers, BLM areas are prized by advocates of wild rivers, wildlife, and wilderness; by anthropologists and historians; by lumber, oil, and gas companies; and by recreation buffs ranging from bird-watchers to dirt bikers. The stakes are high. BLM public lands include more than 270 million acres—one-eighth of the entire U.S. land mass—as well as subsurface resources underlying an additional 300 million acres.

Created in 1946, when the federal General Land Office and the federal Grazing Service merged, the BLM inherited mostly arid and rocky terrain unsuitable for farming and thus passed up by settlers in the land-office giveaways of the last century. Along with these lands, mostly located in ten western states and Alaska, the agency also inherited lax management and a lack of respect. These factors long obscured the cultural and wilderness treasures and little-known spectacular wildlife and scenic values within BLM lands.

Even while they benefited from BLM's admittedly lenient grazing and mining policies, beneficiaries still found time to deride the agency. "Nothing delights the heart of a Nevada cowpoke more than to smell the hide of a BLM director roasting over a sagebrush fire," said one such individual in testimony before Congress, as related in a *Smithsonian* article on BLM (September 1990). Some environmentalists have admitted to similar fantasies.

In 1976, thirty years after the agency's creation, Congress passed the BLM's first charter, the Federal Land Policy and Management Act. It requires that land-management decisions be made in accord with even-handed, multiple-use, and sustained-yield principles while protecting environmental, cultural, and scenic resources.

More than a decade later, however, a September 1989 article in *Sierra* reported that BLM lands were still in poor shape from overgrazing and other abuses: "That the BLM has failed to implement its charter is generally acknowledged by public-lands activists, who recognize that a lack of financial resources accounts for much of the problem. The agency administers one-third more land than the Forest Service does but works with only about half the budget and one-third the staff."

The tide began to turn that same year toward a more balanced approach to multiple-use management. With increased public interest, the once-lethargic BLM has been working on its image and has implemented substantive new programs. High-visibility programs such as Adventures in the Past, Backcountry Byways, and Watchable Wildlife have brought the public into closer contact with BLM's cultural, scenic, and wildlife resources. Controversy continues over the BLM's role in the spotted owl dispute and in leasing of grazing rights and over specific wilderness areas, but the consensus is that wildlife and recreation have achieved a new respect at BLM.

Now the BLM's secret is out: The agency manages the federal government's largest, most varied, and scientifically most important body of cultural resources, with cultural properties estimated in excess of 4 million. BLM also manages sixty-six wilderness areas, covering more than 1.6 million acres in nine western states, and 50 million acres of forested lands. It is charged with managing and protecting fifty thousand free-roaming horses and burros in ten states.

BLM has been adding new employees even as hiring by other agencies has slowed down, going from 7,800 employees in 1989 to 10,000 in 1992. There will be a lot of opportunity at BLM during the next few years as many employees reach retirement age. To meet the challenge of managing millions of acres

Kate Kitchell, Bureau of Land Management
Earth Worker

Kate Kitchell's most essential advice for a conservation career: "You must always remember to keep a sense of humor and believe in yourself." In her present role as resource area manager of Kremmling Resource Area in Colorado, Kate oversees the multiple management of four hundred thousand acres. She enjoys her job and working with the BLM very much. "It [BLM] offers great opportunity for career development."

Kate, 35, says that she has always had an interest in conservation, but her four SCA experiences certainly helped in directing her career. "I always wanted to get into the field, but my SCA experiences provided an opportunity to build contacts and help focus my career direction. SCA definitely provided me with the foundation to grow in my career."

Kate's experience started as an SCA High School Program participant at Canyonlands National Park in Utah in 1975. The following year, she worked as a resource assistant at Assateague Island National Seashore in Maryland. And in the fall of 1977, she worked as
continued

of public lands, BLM relies on the expertise of professional staff from a number of disciplines. Here's a broad sampling of current positions:

ENGINEERS AND TECHNICAL PROFESSIONALS. Civil, agricultural, and electrical engineers; computer specialists and cadastral (land) surveyors; fire management specialists; surface-protection specialists; land use, regional, and urban planners; and environmental specialists.

LAND AND RENEWABLE RESOURCE PROFESSIONALS. Wildlife and fisheries biologists; foresters; range conservationists; hydrologists; specialists in air quality and watersheds, natural resources, recreation, and realty; meteorologists; landscape architects; archaeologists; historians; botanists; ecologists; and planners.

MINERALS PROFESSIONALS. Geologists, mining engineers, petroleum engineers and technicians, physical scientists, mineral economists, hydrologists, leasing specialists, adjudicators, land-law examiners, appraisers, data and/or modeling specialists, and surface-protection specialists.

BLM RANGERS. Unlike the Park Service, BLM has not removed prerequisite educational requirements, so BLM rangers have more training than park rangers do and also are classified in a professional series, bringing greater pay. For more information, contact

Employment Office, U.S. Bureau of Land Management,
Department of the Interior, Washington, DC 20240.

continued from page 77

a co-crew leader at Canyonlands and as a resource assistant at Zion National Park in Utah. In 1979, Kate graduated with a bachelor's in both botany and environmental conservation. After graduation Kate returned to the Park Service and attended Utah State University, receiving her master's degree in recreation resources management.

Several years ago, Kate moved to the Bureau of Land Management. Kate felt she could not move beyond where she was with the Park Service. "I felt as though I had grown as much as I could with the Park Service," says Kate. "I needed to diversify my experience, and BLM offered me the opportunity to do that."

NATIONAL MARINE FISHERIES SERVICE

The National Marine Fisheries Service (NMFS) is a small agency with a large mission. Young compared with the Forest Service or the National Park Service, the agency and its nearly twenty-five hundred men and women are charged with nothing less than the well-being of the living resources of several million square miles of ocean along America's shores—at least a thousand square miles per employee. As the recognition of the earth as a marine habitat grows, the stature of the agency also will grow. For those who qualify, the agency offers a singular opportunity to excel.

Established in 1970 as part of the new National Oceanic and Atmospheric Administration (NOAA), NMFS took marine commercial and sport fishing functions and jurisdiction over most marine mammals and other endangered species from what is now called the U.S. Fish and Wildlife Service and combined these responsibilities with research activities. Its mission greatly expanded under the 1976 extension of the U.S. Exclusive Economic Zone (EEZ) to two hundred nautical miles from shore. In that vast oceanic expanse, NMFS is challenged with promoting the conservation and management of living marine resources that are part of a global web of marine life that knows no boundaries.

In its early years, NMFS set about its task during a boom-time for marine and coastal policy, a decade that featured the Law of the Sea Conference, the Year of the Coast, and passage of laws such as the Magnuson Fishery Conservation and Management Act (which extended the EEZ); the Coastal Zone Management Act; Marine Mammal Protection Act; and Marine Research, Protection, and Sanctuaries Act. Despite the expanded scope assigned NMFS by these statutes, however, funding and staffing for the agency was flat from the late 1970s until the past two years. The current workforce is stretched thin in its task of protection and management of resources, including

- Fish, shellfish, and continental shelf–dwelling species within the EEZ
- Anadromous fish species of U.S. origin throughout their range, except within the EEZ of a foreign nation
- All species of marine mammals except sea otters, polar bears, manatees, and walruses that occur in the U.S. territory or the EEZ
- Species of marine mammals, fish, and sea turtles determined to be endangered or threatened with extinction, currently numbering twenty-three species
- Habitat associated with the species under its jurisdiction

NMFS programs provide support for fisheries management and development, regulation of fisheries, international fisheries affairs, conservation of protected species and habitat, enforcement under various marine mammal protection and endangered species laws, and scientific research.

MANAGING FISHERIES

As a manager of the EEZ, NMFS is responsible for more than 2 million square miles of ocean. States are responsible for managing fisheries within their territorial waters, which in most cases extend three miles from shore. From these state and federal waters are taken about one-sixth of the entire ocean's annual fish and shellfish harvest. More than three hundred species are harvested commercially.

The extension of U.S. jurisdiction to two hundred miles from shore led to increased harvesting of U.S. fishery resources by American fleets. Each year, more and more finfish and shellfish species are harvested by U.S. commercial and recreational fishermen. In addition, many valuable species that migrate through and beyond our two-hundred-mile zone—such as tuna, swordfish, and salmon—are also heavily fished by other nations.

Directly or indirectly, U.S. commercial fisheries contribute $50 billion annually to the nation's economy. These marine and coastal resources also support aquaculture, recreational diving, and Native American subsistence. While 93,000 commercial fishing craft ply U.S. waters annually, an additional 17 million Americans enjoy recreational saltwater fishing and catch about one-fifth of the total food harvest.

The National Marine Fisheries Service, together with eight Regional Fishery Management Councils and the coastal states, manages U.S. marine fisheries under the authority of the Magnuson Fishery Conservation and Management Act, the Fish and Wildlife Coordination Act, and many other federal laws. To develop fishery management plans for all species under federal control, NMFS regional and headquarters personnel work with the Fishery Management Councils, which are composed of representatives of state governments, commercial and recreational fisheries, and environmental groups.

PROTECTING IMPERILED SPECIES

Under the Endangered Species Act and the Marine Mammal Protection Act, NMFS serves as steward of many protected marine species. Some species of dolphins, whales, seals, sea lions, and sea turtles and some stocks of Pacific salmon have declined in abundance so much that their future existence is now in jeopardy. The agency is charged not only with protecting these living resources from harm but also with protecting critical habitat for them.

PRESERVING AND RESTORING HABITAT

NMFS monitors and protects the nation's coastal habitats—estuarine marshes, coral reefs, sea grass beds, and mangrove stands—whose health is vitally important to living marine resources. The alarming loss of wetlands due to development, pollution, subsidence, and dredging and filling seriously jeopardizes the productivity of U.S. fisheries. NMFS shares in the management of NOAA's Damage Assessment and Restoration Program, which works to mitigate coastal habitat damage resulting from oil and chemical spills and other environmental disasters.

Nina Garfield, National Oceanic and Atmospheric Administration
Earth Worker

Nina Garfield, a 30-year-old Pittsburgh native, says, "I've always been interested in the sea. We went to Cape May, New Jersey, every summer when I was child, and that love of the ocean has been in me since then."

A degree in sociology with a psychology minor from Kalamazoo College in Michigan prepared her for three years of work in a psychiatric hospital as a research assistant. But she eventually decided to study what she was most passionate about—the ocean.

While volunteering at the Pittsburgh Aquarium, she learned about the fisheries management field from the aquarium director and felt that it was a field that could combine her sociology background with her love for the sea.

She studied at the University of Rhode Island. Diving one summer at a National Marine Fisheries Service laboratory in Galveston, Texas, she began to develop her thesis about the social impacts of the closing of a shrimp fishery in Texas.

continued

INSPECTING SEAFOOD

NMFS plays a key role in safeguarding the health of consumers who enjoy seafood. Together with the Food and Drug Administration, NMFS inspects hundreds of processing plants, distributors, and vessels and works cooperatively with other nations to ensure that both domestic and imported fish and shellfish are safe to eat.

SERVING NATIONWIDE

Headquartered in Silver Spring, Maryland, near the nation's capital, the NMFS carries out its duties in five regions: Northeast, Southeast (including the U.S. Caribbean), Southwest (including Hawaii and U.S. South Pacific territories), Northwest, and Alaska.

Most NMFS employees work in regional offices and other field facilities. Each regional office is served by a Science and Research Center where scientists conduct the studies necessary to support management decisions. They also work at twenty-four NMFS laboratories that contribute to this important work, collecting fisheries statistics, conducting resource and environmental surveys, studying the biology and population structures of marine species, analyzing marine and coastal ecosystems, and investigating contamination of seafood.

The NMFS science centers work closely with other NOAA units—the National Ocean Service, Office of Oceanic and Atmospheric Research, National Weather Service, NOAA Corps, Coastal Ocean Program, Climate and Global Change Program, and National Environmental Satellite Data and Information Service—to carry out NOAA's mission of monitoring, recording, and predicting changes in the oceans and atmosphere.

WORKING INTERNATIONALLY

NMFS fishery managers and scientists work with colleagues from around the world under international treaties and agreements to conserve marine animals shared with other nations. They are key participants in organizations such as the International Whaling Commission, International Commission for the Conservation of

continued from page 80

Determined to learn how the federal government operated, Garfield decided to get a job in Washington, D.C. Soon, she was offered a position as a program specialist with the sanctuary and reserve division within NOAA's Office of Coastal Resource Management, where she is one of two people specifically working on establishing the national sanctuary on the Olympic Coast mandated by Congress through the Marine Protection, Research, and Sanctuaries Act.

Today, she regrets that she did not pursue a stronger science curriculum and is considering a master's program in either ecology or conservation biology.

For those considering a career in the field, she concludes, "My biggest suggestion is to blend science and management. I knew people in oceanography graduate school who did not know that there were management zones in the ocean. That seemed incredible to me. Today, I know people in management who cannot speak in scientists' language and vice versa. They don't understand the compromises that need to be taken into account to help manage our oceans and shorelines."

Atlantic Tunas, Inter-American Tropical Tuna Commission, International Commission for Exploration of the Seas, Pacific Halibut Commission, and the Convention for the Conservation of Antarctic Marine Living Resources.

REPRESENTATIVE CAREERS IN NMFS

Occupations at NMFS are divided into four fields—data management and operations support; statistics; clerical, administrative, and technical support; and research coordination. Under Field IV, research coordination, job titles include biological aide, fishery management specialist, biological technician, fishery biologist, fishery administrator, operations research analyst, and co-op student trainee.

NOAA's fishery biologists, who are found primarily in its NMFS, work at various research facilities throughout the United States. They study the life history of fish and shellfish, attack problems of disease, and identify and study oceanic stocks of fish. They also study the effects of environmental and human-induced changes on fish, regulate fisheries to restore depleted resources, and develop conservation gear and fishing practices to ensure a continuing, maximum yield. In addition, they participate in implementation of international treaties to conserve and maintain the oceanic resources. Specialization is possible in such fields as oceanography, population biology, and ecology.

The job requirements for a fishery biologist position (series 0482) may be designated as general, management, or research, depending on the type of position. The NOAA fishery biologist career ladder starts at GS-5/7 and climbs to GS-15. About 12 percent of the positions are at the GS-5/7 level; more than half of all fishery biologists at the agency are at the GS-12 to GS-14 levels. Entry-level GS-5 positions require completion of all requirements for a bachelor's or higher degree with a major in the biological or agricultural sciences or completion of the appropriate course requirements (request a list) plus an equivalent amount of experience and education. Additional experience or graduate degrees qualify applicants for higher grades.

Fishery biologists may advance into fishery administrator or other management positions. Fishery administrators (GS 13/14), for example, are responsible for planning, directing, and organizing management and conservation programs, marine mammal and protected species programs, trade analysis, fishery statistics, fishery monitoring programs, and market news. They also administer programs to analyze trade and economic conditions and may be involved in developing regulations on domestic and foreign fishing.

NMFS also hires from other series in a variety of disciplines—such as wildlife biologist, ecologist, hydraulic engineer, and chemist. Fishery management specialists (GS-301), for instance, do not need a degree in biology as long as they have some fisheries experience, because they concentrate on the economic and social aspects of management. Staff who work on treaties and legislation may have degrees in law, political science, or international affairs. Fishery reporting specialists have career potential ranging from GS-5 through GS-12 and can begin with a degree in any field leading to a bachelor's degree.

Some opportunities are also available for those without college degrees. For example, biological technicians (GS-4 through GS-6) serve as part of a cooperative industry-federal program created to monitor injury and killing of marine mammals in the course of net fishing.

The observer program is a good area for people starting their careers. For example, the government has invested a lot of money since the 1970s in the problem of the catching of dolphins in tuna nets. The NMFS has regulatory authority under the Marine Mammal Protection Act for dolphin mortality observers on tuna boats. These observers helped document the capture of dolphins, leading to more tuna companies implementing dolphin-safe tuna fishing methods. These observers were federal employees, although the use of observers now varies around the country, with private firms receiving observer contracts for various fisheries in some areas. Many of these jobs are in the Pacific Northwest and Alaska.

Although young people can find positions in observer programs, at regional science centers, and in collection of statistics, by and large, NMFS tries to hire Ph.D.-level scientists.

For information about employment with NMFS and other National Oceanic and Atmospheric Administration (NOAA) jobs, contact offices in Seattle, Washington; Boulder, Colorado; Kansas City, Missouri; or Norfolk, Virginia, or write

> *NOAA, Office of Administration, Personnel Operations Division, Staffing and Employee Development Branch, 1335 East-West Highway, Silver Spring, MD 20910, (301) 713-0677.*

U.S. ENVIRONMENTAL PROTECTION AGENCY

The U.S. Environmental Protection Agency is the federal agency charged with implementing and enforcing federal laws created to protect the environment. EPA was molded from a consolidation of the environmental operations of fifteen environmental offices in five other executive departments and independent agencies. The agency is relatively young—it began operations on December 2, 1970.

EPA's original job was regulating air pollution control, solid waste management, radiation control, and the drinking water program. The accelerated increase of stringent environmental laws has broadened the EPA mission into other areas, such as pollution prevention, hazardous waste, toxic waste, environmental impact review, wetlands, underground storage tanks, spill response, and other environmental issues. What EPA does affects and benefits every American and influences our global community. Although the U.S. Office of Personnel Management (OPM) predicts the total number of federal employees is unlikely to grow significantly in coming years, EPA was still experiencing a steady increase in its workforce, from five thousand employees at its inception in 1970 to almost four times that in 1992. However, a former director of EPA's National Recruitment Center said the agency may cut some personnel during the "reinventing government" initiative of the Clinton administration. Nevertheless, the administration's strong stand on environmental protection may help EPA avoid serious staffing limitations relative to other agencies.

About eight thousand of EPA's employees work at the headquarters in Washington, D.C. Other employees work in the regional offices, laboratories, or field stations. Regional offices are located in Atlanta, Boston, Chicago, Dallas, Denver, Kansas City, New York, Philadelphia, San Francisco, and Seattle.

Major research and development laboratories are located in Research Triangle Park (North Carolina), Cincinnati, and Las Vegas. Other research laboratories are located throughout the United States, including Ada (Oklahoma), Athens (Georgia), Corvallis (Oregon), Duluth, Edison (New Jersey), Grosse Isle (Michigan), Gulf Breeze (Florida), Monticello (Minnesota), Narragansett (Rhode Island), Newport (Oregon), and Warrenton (Virginia). EPA field offices are located in Ann Arbor (Michigan), and Montgomery (Alabama). The EPA National Enforcement Investigations Center is found in Denver. EPA employees are stationed in thirty-seven states and eight countries and territories worldwide.

Prospective technical employees can find work in various environmental programs, including

> *Air and Radiation*
> *Water*
> *Prevention, Pesticides, and Toxic Substances*
> *Solid Waste and Emergency Response*
> *International Activities*
> *Communication, Education, and Public Affairs*
> *Policy, Planning, and Evaluation*
> *Enforcement*

AIR AND RADIATION

The Office of Air and Radiation (OAR) manages indoor and outdoor air quality, ozone, acid rain, radon, radioactive wastes, nuclear accident response, and related training. Many OAR employees have their work cut out for them with creating and implementing regulations under the 1990 Clean Air Act Amendments (CAAA), which significantly strengthened the air quality law. Technical OAR employees include engineers (chemical, environmental, and mechanical), meteorologists, environmental scientists, and health physicists.

WATER

Restoring, protecting, and enhancing drinking water, fresh water, marine and estuarine water, wetlands, and groundwater are the basis of environmental careers in EPA's Office of Water. Work in a water quality career can vary from developing interagency and international management policy for freshwater ecosystems in the Great Lakes to setting critical drinking water quality standards to challenging the Army Corps of Engineers' decisions on wetlands permits. EPA's Office of Water hires biologists, chemists, environmental engineers, environmental protection specialists, environmental scientists, geologists, hydrologists, and toxicologists.

According to one Office of Water supervisor, drinking water and watershed management are the "new wave" of growth in the water program area. He cites EPA documents that highlight drinking water and watershed management as high priority EPA programs and budgets programmed for $1 billion per year, until the year 2000, for a drinking water state revolving fund.

PREVENTION, PESTICIDES, AND TOXIC SUBSTANCES

The Office of Prevention, Pesticides, and Toxic Substances (OPPTS) is the type of organization inspired by Rachel Carson's classic book *Silent Spring*. The book that started the modern environmental movement focused on the federal government's inconsistent and inadequate pesticide and toxic substance management, application standards, and ecological monitoring. EPA's OPPTS is designed to consistently use the best science to reduce use of pesticides and toxic substances and to set standards to properly apply, manage, and monitor any use of these substances and to enforce penalties for misuse. This element of EPA has the very difficult task of evaluating risk and benefit of various chemicals and developing appropriate restrictions.

Typical career tracks include those for chemical engineers, biologists, environmental engineers, environmental scientists, pharmacologists, environmental protection specialists, toxicologists, and economists.

OFFICE OF SOLID WASTE AND EMERGENCY RESPONSE

Some of the greatest workload increases within EPA over the past ten years were probably experienced by the Office of Solid Waste and Emergency Response. Reauthorization of the Resource Conservation and Recovery Act (RCRA) in 1984 and passage of the Superfund Amendments Reauthorization Act (SARA) of 1986 created a renewed national push to prevent pollution and clean up contaminated sites, and EPA's Office of Solid Waste and Emergency Response (OSWER) responded. OSWER is charged with developing EPA policy; developing guidance and standards; and providing technical assistance in solid waste management, hazardous waste management, underground storage tank management, emergency spill response, and contaminated sites cleanup.

Private environmental personnel headhunters and environmental consultants often seek experienced scientists and engineers from this office. There is a very strong outside-employer market for scientists and engineers who have field experience in site assessments and remediation and know both the regulations and the regulator, EPA, well. EPA's OSWER makes extra efforts to attract and retain talented scientists and engineers who have a strong commitment to public service in pollution prevention and contamination cleanup.

One environmental specialist in EPA's underground storage tank program likes the access staff members have to senior managers as well as their flexibility to work on a wide variety of projects. He sees the underground storage tank program growing, with a peak prior to the year 2000. His advice to job seekers is to meet with EPA supervisors and express interest in working for them.

OSWER needs environmental scientists, chemists, hydrogeologists, engineers, physical scientists, and other technical personnel.

INTERNATIONAL ACTIVITIES

Environmental problems are global concerns. EPA's Office of International Activities works with other countries to solve ecological problems such as ozone depletion, global warming, international hazardous waste management, and protection of marine and polar environments and provides technical assistance to developing countries.

What an opportunity to help the environment! As an OIA employee you could play a major role in influencing international action to solve environmental problems. You could be involved in developing international environmental agreements, monitoring compliance, and providing environmental protection assistance to other countries as the political climate allows.

Christiane Blume, Environmental Protection Agency
Earth Worker

Christiane Saada Blume, an EPA official, says that she has known since elementary school that she wanted to be involved in environmental protection. That is one of the reasons she enjoyed working as an SCA resource assistant during the summer of 1982 at Kootenai National Forest in Montana. Christiane assisted a forest biologist in range surveys of elk and deer. "It was a great conservation experience," she says.

Christiane received her B.S. in engineering and management systems from Princeton University, and her M.S. in environmental engineering from Illinois Institute of Technology. Shortly after completing graduate studies, she began working for EPA.

When she first joined the EPA, Christiane worked with the Office of Groundwater Programs, which deals with groundwater protection issues. In 1988 she received the EPA Bronze Medal for her efforts in developing the Groundwater Protection Strategy for EPA Region 5. At present she is chief of the State

continued

Of course, Peace Corps alumni are excellent candidates for a career in this office. Peace Corps alumni gain over two years' international experience, and EPA may hire them directly without going through other OPM hiring procedures.

EPA's OIA seeks persons with backgrounds in international studies, international relations, economics, political science, natural resources, and environmental science. The Clinton administration plans to hire environmental specialists at U.S. embassies abroad to help export American environmental technology products and services to other countries and increase jobs at home. Employees from EPA's OIA would play a major role in that federal "eco-preneurial" effort.

continued from page 86

Programs Unit for the Drinking Water Section.

Serving as a liaison to states and the federal government, Christiane helps to administer drinking water programs that ensure that public water supplies follow the Safe Drinking Water Act regulations. What Christiane likes best about her position is, "What I'm doing is helping to protect the health of the public and that's important."

RESEARCH AND DEVELOPMENT

If you want a job in research, EPA has opportunities for you in the Office of Research and Development (ORD) and research laboratories. According to one researcher at a laboratory in Minnesota, the best thing about working for EPA is freedom to do real research that contributes to protecting our environment.

ORD's mission is to determine the impact of pollutants on the ecosystem and human health. ORD laboratory research and development cuts across all EPA resource areas, including air, water, hazardous waste, solid waste, groundwater, contaminated soil, climate, and ecosystems. Research and development focuses on five program areas, including

> *Health effects*
> *Environmental engineering technology*
> *Environmental processes and effects*
> *Monitoring and quality assurance*
> *Health and environmental assessment*

The EPA educational profile shows that about 68 percent of employees have college degrees, 20 percent have master's degrees, and 6 percent have doctorates. EPA's research and development laboratories prefer individuals with graduate degrees and hire most of EPA's doctoral employees.

JOB HUNTING STRATEGIES

Prospective environmental employees may enter EPA through the standard hiring procedures discussed under "Federal Hiring Procedures" earlier in this chapter, including OPM

BEST SOURCES OF EPA JOB VACANCY INFORMATION

The Environmental Career Center asked EPA personnel specialists, "What is the best way to keep updated on job vacancies at your office and offices you serve?" EPA's National Recruitment Center, each regional office, and the three major laboratory centers in Research Triangle Park, Las Vegas, and Cincinnati were contacted. Eight of fourteen offices have recorded job "hot lines" you can call. Those without job hot lines suggest you call them periodically or call the nearest Federal Job Information Center.

EPA Office	Jobs Hot Line	Personnel Office
Boston	No	(617) 565-3719
New York	No	(212) 264-0016
Philadelphia	No	(215) 597-8922
Atlanta	(800) 833-8130	(404) 347-3486
Chicago	No	(312) 353-2027
Dallas	(214) 655-6444	(214) 655-6444
Kansas City	(913) 551-7068	(913) 551-7041
Denver	(303) 293-1564	(303) 293-1499
San Francisco	(415) 744-1111	(415) 744-1300
Seattle	(206) 553-1240	(206) 553-2959
Washington, D.C.	No	(202) 260-3144
Research Triangle Park	(919) 541-3014	(919) 541-3071
Las Vegas*	No	(702) 798-2402
Cincinnati	(513) 569-7840	(513) 569-7801

For more information about EPA careers, contact

U.S. Environmental Protection Agency, National Recruitment Center, 401 M Street, Washington, DC 20460, (202) 260-3144.

* Provides staffing services for research laboratories nationwide.

registers, the Outstanding Scholar program, cooperative education programs, the Presidential Management Intern program, the Stay-in-School program, and Peace Corps alumni eligibility.

EPA MANAGEMENT INTERN PROGRAM

The EPA Management Intern Program (EMIP) is an excellent program for highly motivated individuals of diverse backgrounds who are near completion of an undergraduate or graduate degree. It is a two-year program, with special training and benefits, that can lead to a permanent position with EPA. Rotational assignments among various EPA facilities and localities provide EMIP participants outstanding cross-training—these interns truly learn the broad EPA mission in a personal way.

Students should be U.S. citizens with at least a 3.5 GPA overall and have a strong commitment to public service. Application deadlines are usually at the end of January, and you should contact the EMIP coordinator at the EPA headquarters for application details.

VOLUNTEERS

Volunteering is another strategy that works. EPA will remember you when a vacancy opens if you can sacrifice several hours or more per week to help manage EPA's tremendous workload. Even if an EPA job does not materialize as soon as you want, you will at least gain professional experience within an agency that is the global leader in environmental protection. That is an excellent résumé eye-catcher for other prospective employers.

Chapter Six

WORKING FOR NONPROFIT ORGANIZATIONS

As we enter the twenty-first century, nonprofit conservation and environmental protection organizations in the United States are almost as numerous and varied as the kinds of ecosystems in North America. According to the Conservation Fund, the more than ten thousand nonprofit environmental groups in the United States range from traditional land and wildlife conservation or nature education groups to hybrids of health and pollution control organizations.

These organizations vary in size from one-person offices to staffs of several thousand in offices and field centers around the nation. In addition to biologists, foresters, and other natural resource professionals, nonprofits also employ educators, managers, lobbyists, lawyers, writers, public relations and human resources specialists, economists, fundraisers, accountants, membership specialists, and a variety of other professionals.

Some groups are mainly advocacy organizations; others serve primarily as think tanks; still others focus on public environmental education. Most have evolved over the years into some combination of the preceding. The labels given these groups range from "liberal" to "middle-of-the-road" to "conservative," even though most are nonpartisan. Some were founded on a philosophy of preserving wilderness, whereas others trace their beginnings to foresters interested in multiple uses of land.

Therefore, instead of seeing conservation organizations as streams stemming from the same river of conservation philosophy, it is helpful to see them as diverse, living organisms that interrelate in what may be viewed as the most "live" social ecosystem of the twentieth century—environmentalism. Just as scientists are promoting a new appreciation of biological diversity, the nonprofit world needs to develop its own social and cultural diversity. Not all organizations will appeal to you equally, but each plays a role in relation to the others.

ECOLOGICAL ROLES FOR ENVIRONMENTAL GROUPS

Your research will convince you that not every organization suits your style, but nonetheless it is important to remember that each group plays an "ecological" role in the conservation movement. Whether you see yourself on the left or right or in the middle, remember that how each advocacy organization positions itself on an issue helps define the others. For example, when then Secretary of the Interior James Watt raised a ruckus in the early 1980s with calls for opening wilderness areas to development, groups seen as more liberal—such as The Wilderness Society and the Sierra Club—led the charge against him. Their actions defined the issues for Congress and the media and set the stage for other nonprofit organizations. The more conservative National Wildlife Federation did not react as quickly, but when it did, it had enormous impact because most of its members had voted for President Reagan and because it boasted the largest membership.

The memberships of organizations help determine their positions. The National Wildlife Federation has many hunters as members, for example, but the Humane Society of the United States does not. Both organizations have large conservation programs and members who are strong conservationists, but because of their different constituencies, the two organizations do not always agree on the issues.

FINDING THE RIGHT NONPROFIT FOR YOU

To find which organizations interest you, do research. Consult the reference books in the Recommended Book List at the back of this book, or write to some of the nonprofit employers listed at the end of this chapter. This list includes mostly larger organizations, which have openings more frequently; the directory at the back of the book lists all *Earth Work* job listers—it will give you an idea of the range of organizations with jobs in the nonprofit and other sectors. You will also find the *Conservation Directory* of the National Wildlife Federation a useful source on nonprofit organizations because all the descriptions were submitted by the organizations themselves.

Study the magazines of conservation organizations at your local library, and request copies of their annual reports to get an idea of their policies. Use an inexpensive computer network such as Econet to keep up with current developments. Although *Earth Work* focuses in this chapter on a sampling of larger nonprofit organizations, don't forget that much of the best work of conservation organizations is done at the grassroots level—by local and regional nonprofits and chapters of the national organizations. The *Conservation Directory* features a state-by-state breakdown of them. In addition to advocacy groups, also research professional associations and educational institutions that operate as nonprofits. The *Earth Journal Environmental Almanac and Resource Directory* features a guide to outdoor education centers.

Keep in mind that small and midsized organizations have relatively few job openings and receive tons of unsolicited mail from prospective employees, mail that they often do not have enough staff to handle suitably. (If you are writing for information rather than in response to a particular job opening, your chances of receiving a reply to your letter will increase exponentially if you volunteer to help with the mail!) On the other hand, you often have opportunity for broader responsibilities and initiative at these smaller groups.

For example, in starting a new job at the Environmental Investigation Agency, Courtney Stark, a law student, is not only lobbying on international trade in endangered species such as tigers but also helping to set up press procedures at the organization. Because the job is "just what I wanted to do after law school," she is delaying completion of law school to take it. Courtney is just one of many who have been offered full-time paid jobs after performing volunteer work or internships at nonprofits. If you are looking for that all-important first job, indicate whether you are open to an internship and if so, whether you need a stipend. If you have a lot of experience, do your research before writing, and explain up front how you can be of specific assistance to that organization.

To keep abreast of which organizations have job openings, watch *Earth Work* magazine's job listings carefully each month; they are the only listing published from within the nonprofit conservation community, and the publisher, the Student Conservation Association, is a nonprofit. The listings include positions at all experience levels and allow you to compare nonprofit jobs with those in the government and private, for-profit sectors.

Steve McCormick, The Nature Conservancy
Earth Worker

Steve McCormick, Director of The Nature Conservancy's California Region, has always had an interest in conservation. A participant in the Student Conservation Association High School Program in 1968 at Olympic National Park, Steve went on to the University of California at Berkeley and received his bachelor's in natural resource economics in 1973. After graduating from Hastings College of Law in San Francisco in 1976, Steve went to work for the California Coastal Commission as a lawyer and permit analyst. After two years at the commission, he left to join the ranks of The Nature Conservancy, a national membership organization committed to preserving biological diversity through land acquisition and preservation. "When I was in college I felt that a background in law would be indispensable," he says, "but I didn't want a traditional legal career."

More than fifteen years later, Steve is still with The Nature Conservancy and is certainly not practicing a traditional legal career. Much has changed since he joined it in

continued

You will notice that the larger nonprofits such as The Nature Conservancy (TNC) have job listings every month in the magazine. TNC and the World Wildlife Fund have their own job hot lines as well.

If you choose to pursue a career with one of these organizations, you will have a large enough selection of nonprofits to readily find one that matches your own political beliefs and work aspirations. For example, if you want a nonprofit career promoting wildlife conservation, you have a number of choices: a national advocacy group that champions wildlife in addition to other environmental causes, such as the National Audubon Society or the Sierra Club, which have long-standing conservation programs; the larger and more conservative National Wildlife Federation, which has the most members of any conservation organization and impressive educational programs; or Defenders of Wildlife, which zeros in on the protection of wildlife and overall biodiversity and packs a big punch in Congress and the courts with a midsized staff and a "progressive" membership.

If you are seeking an overseas job or one with a distinctly international focus, you might prefer the World Wildlife Fund. If you lean more to a wildlife management–oriented group in tune with the hunting community, you would select a group such as the Wildlife Management Institute. Or, if you would rather focus your efforts on particular species or types of habitats, you will find a range of organizations with specific foci ranging from the American Cetacean Society to the Rainforest Action Network.

On the other hand, if you prefer research to lobbying or field work and have an advanced degree, your application would be considered by policy "think tanks" such as the World Resources Institute. Instead of examining conservation policies, you might prefer to help directly acquire or manage specific wildlife habitats in a position with a land trust or conservancy such as The Nature Conservancy.

Whether you are a lawyer, a scientist, or a publicist, you can help protect wildlife or participate in other conservation and environmental efforts. What is more, you can choose a variety of ways to do so, from strategizing policy at a conservation organization to strategizing propagation at one of the progressive zoological parks. You might even be lucky enough to do both during your career.

continued from page 92

1978, from one office for thirteen western states to at least ten offices today in California alone. He thoroughly enjoys his work with TNC and takes pride in his involvement in TNC's significant growth in California. He respects and enjoys "the scientific and objective way TNC decides its priorities and the scientific and objective way it works with other organizations."

As director of TNC's California Region, he manages land acquisitions and stewardship, raises funds, coordinates all California regional activities, and oversees the administration of TNC's California offices.

Steve served on the board of directors of the Student Conservation Association from 1982 to 1988, a position that brought his job and volunteer work together in California: "While on the board I got to see SCA high school work groups in California. To see those kids working out in the field is an uplifting experience." To those interested in pursuing a career in conservation, Steve suggests, "Look around and find out what's out there, who does what, what fits for you. If you can't find a job straight out of school, do volunteer work."

Richard Block, senior fellow and former director of public programs at the World Wildlife Fund (WWF), initially applied his master's in natural resources from the University of Michigan to work at engineering firms and then to teaching at the university's School of Natural Resources. Next he directed public relations for the Kansas City Zoo and educational programs for Zoo Atlanta. In 1987 he went to WWF to build public education programs and currently is expanding the organization's marketing and licensing activities.

Richard says, "It's been a wonderful opportunity to work in a variety of fields, including everything from policy issues to field work with bugs and bunnies to education to marketing the environment."

THE WORLD OF ADVOCACY

Nonprofit organizations afford you an opportunity not only to work for a good cause but also to be an outspoken *advocate* for it in ways that would be impossible in the government or for-profit sectors. No other kind of organization gives you such freedom to pursue your ideals—or so much work—as a nonprofit does.

Christopher Croft entered the nonprofit world in 1988 after he was asked to resign his job as a federal biologist and tuna boat observer following publication of an op-ed piece on the need for labeling tuna fish cans. Croft wrote that the cans should be labeled to inform consumers whether the company had fished for the tuna in a "dolphin-safe" manner. Chris worked for Greenpeace and founded Environmental Solutions International before going to Defenders of Wildlife as marine wildlife coordinator. At Defenders he created the Dolphin Coalition, an alliance of thirty-seven national and international organizations, to win passage of the Dolphin Protection Consumer Information Act of 1990. Chris explains, "It's important for people to examine their lives. Do you want to carry out policies or initiate them? For me, it's been worth the risks to help make policy."

PREPARATION FOR NONPROFIT CAREERS

Like Croft, many conservationists have entered careers in environmental advocacy with a scientific background. Others entered conservation or environmental protection careers with backgrounds in political science or communications—useful skills, considering that environmental advocacy involves translating scientific and policy information for the general public and decision makers.

Margaret Podlich, who received a B.A. in history from Tulane University, considers herself one such environmental "translator." Although she sometimes feels constrained by her lack of a scientific background, Margaret realized on her first day on the job at the Chesapeake Bay Foundation in Maryland that her liberal arts skills would be useful in public communications. The grassroots department was preparing an action alert scheduled to go out to citizen activists. Podlich says, "The topic was fish passage, and I stumbled on the word *anadromous.* The writers of the draft, both trained in science, had not considered that their audience, average citizens, might not know what the word meant," she recalls. The draft was revised, and the experience taught her "to translate scientific information into words that nonscientific people could understand and then use to help the environment." Podlich used her communications skills in outreach and education programs teaching the public about the impacts of human beings on the Chesapeake Bay system. After four years with this regional nonprofit conservation group, she took a position with the Center for Marine Conservation in Washington, D.C., where she could continue to do outreach and educational work about marine and coastal pollution issues on the national level.

Peter Harnik, who is currently vice president of the Rails to Trails Conservancy, studied to be a journalist, not a conservationist. However, shortly after the first Earth Day in 1970, his writing as assistant editor of *Earth Tool Kit* and as an editor at Environmental Action led him into the policy world. He served as coordinator of Environmental Action from 1973 to 1977. While an editor, Harnik studied the concept of converting unused rail corridors into trails for bikers and hikers. His research helped lead to the founding of Rails to Trails in 1986.

Harnik heartily recommends working for nonprofit organizations to those just embarking on an environmental career. Through his diverse experience, Peter says, "I've learned that individuals can make a tremendous difference if they understand the pressure points in the system, learn how the system operates, and learn timing. . . . It's great to work on an issue that's personally meaningful while creating a heritage for future generations." In addition, the popularity of the environmental movement has improved pay and benefits. "You can make a living doing good things and helping other people," Harnik emphasizes.

Bruce Hamilton, Sierra Club
Earth Worker

Bruce Hamilton, director of conservation field services for the national Sierra Club, has had conservation adventures around the world from New York to New Zealand. Today he oversees more than fifteen regional field offices and their conservation programs for the Sierra Club from Alaska to Florida.

Bruce has always had a strong interest in conservation. As a Student Conservation Association high school volunteer in 1967, he worked in the Merck Forest in Vermont, where he enjoyed the work and the company. He became friends with SCA supervisors Bob and Libby Mills, a "kind Quaker couple from whom I learned much about life and with whom I kept in touch later."

Bruce, the son of a Cornell University professor of forestry and natural resources, went through a period of "rebellion against science." He attended the State University of New York at Old Westbury. In its early years, the college was a testing ground for experimental education; the students determined their own course of

continued

Environmental organizations obviously need not only people with scientific and legal expertise but also those with "liberal arts" skills. Increasingly, however, they need people with a combination of both types of skills. For example, while earning his M.S. in environmental science from the University of Montana, Hank Fischer took more journalism courses than environmental courses. That background has paid off in his work as Rocky Mountain representative of Defenders of Wildlife. Hank, who has worked longer than any other representative of a national group in the region, uses a variety of communications to champion protection of the wolf and the grizzly, and biological diversity in general. He writes alerts, advertisements, and op-ed pieces and speaks publicly at forums ranging from local field hearings to network television and his own radio show. He also travels from congressional hearing rooms to remote forests.

Employees with business backgrounds and fund-raisers are in special demand at nonprofits and command high salaries. In November 1993, for example, The Nature Conservancy listed openings in *Earth Work* for an aids grant specialist, assistant director of corporate support, assistant director of foundation support, director of development, and several administrative assistants.

In a nutshell, you can prepare yourself for a nonprofit job in several ways. First, you can choose one of the new specialized environmental or natural resources majors and get your degree in an area not generally available several decades ago. For example, in addition to new science-based majors, an increasing number of universities offer degrees in environmental journalism. Second, you can obtain a traditional degree in one of the sciences or a good liberal arts education; if you choose the latter, however, include natural resources, environmental, and political science courses. Because competition is often stiff—even for jobs with relatively low pay—an advanced degree or volunteer experience providing on-the-job training can help you get ahead. Languages also give a boost to environmental and conservation careers, which are increasingly international in nature.

Several decades ago, environmentalism and work at nonprofits in particular was not considered a career. People in environmental and natural resources careers initially were defined by their job descriptions—lawyer, lobbyist, writer, biologist, or whatever. Today, you can define yourself as a "career environmentalist" or "green" your present career in a variety of job descriptions.

continued from page 95

study and took Outward Bound–style canoe and camping trips. Bird-watching and mountaineering trips helped reawaken Bruce's interest in natural history.

He studied marine ecology by sailing and living on a boat. For three months he lived on Smith Island, a barrier island off the Delmarva Peninsula in Virginia. The island had been bought by developers, and Bruce was hired to help conduct an ecological survey of the area. Feeling that the island should not be developed, Hamilton contacted The Nature Conservancy and shared his findings. Ultimately the developers sold the land to TNC, which established the Virginia Coastal Island Preserve.

He next lined up a position with Scripps Oceanographic Institute in California. He drove across country and subsequently worked as a marine technician on a research expedition to the South Pacific. But of all the places he explored, he especially savored memories of the mountains of Colorado from his cross-country trip. When the SUNY Old Westbury experimental education program closed down, he enrolled at Colorado State University, where he later graduated summa cum laude

continued

EVOLUTION OF NONPROFIT ADVOCACY GROUPS

I n choosing a nonprofit that fits your expectations, it helps not only to research an organization's current programs but also to know a little about the evolution of the nonprofit conservation movement in general.

Many of these groups were formed in the past decade; others date back to early in the century. The first organizations generally were founded in the late nineteenth century or early twentieth century by people interested in protecting parklands and wildlife. The Sierra Club—founded by John Muir in 1892— already has celebrated its centennial, whereas the National Audubon Society (1905), the National Parks and Conservation Association (1919), The Wilderness Society (1935), and the National Wildlife Federation (1936) date to the early part of this century. Of these groups, the Sierra Club, the National Audubon Society, and the National Wildlife Federation—three of the nation's largest environmental groups—later widened their mandates to include environmental pollution as well as conservation issues.

In fact, after the blossoming of land and wildlife conservation organizations in the first half of the century, the next biggest catalyst for the formation of new environmental organizations was the need to deal with pesticides and air and water pollution, problems that rose to the forefront in the 1960s and 1970s. The formation of the Natural Resources Defense Council (1970), Environmental Defense Fund (1967), and Sierra Club Legal Defense Fund (1971) marked a successful marriage of advocacy and law. The slew of environmental laws passed in the 1970s would need these watchdogs to ensure compliance. Other environmental organizations formed at the time included Greenpeace (1970) and Environmental Action (1970), which was formed by some of the leaders of the first Earth Day.

Philip Shabecoff, former *New York Times* environmental reporter, reports most eloquently on the effects of the first Earth

continued from page 96

with a B.S. in wildlife biology and natural resources administration in 1973.

While in college, Bruce became active in lobbying the Colorado state legislature trying to head off oil shale development. An outlet for raising public awareness about the issue was the *High Country News*, a highly respected regional environmental newspaper. Bruce wrote regular articles for the paper and also did field research in wildlife biology and environmental impact analyses on a consulting basis. After graduation, Bruce and his wife, Joan, moved to Wyoming to assist *High Country News* editor Tom Bell. In less than a year Bell moved to Oregon and gave Bruce and Joan stewardship of the paper, with instructions to keep publishing "until the string ran out." Over time the circulation grew as they churned out incisive articles that influenced conservation policy nationwide.

In 1976, Bruce became Northern Plains regional representative of the Sierra Club. Also a veteran of the club's national campaigns to preserve wildlands in Alaska, strengthen the Clean Air Act, and develop a national energy policy, Bruce moved to the national headquarters of the Sierra Club

continued

Day in his history of the environmental movement, *A Fierce Green Fire.* He says, "April 22, 1970, is as good a date as any to point to the day environmentalism in the United States began to emerge as a mass social movement. The American people, demonstrating the power of a democracy to address a social crisis, started taking matters into their own hands. The time had come to save ourselves."

continued from page 97

in San Francisco a decade ago. Joan is senior editor of the club's acclaimed *Sierra* magazine.

As a member of the Sierra Club's senior conservation managers team, Bruce has been active in planning the next generation of work in protecting North America's "critical ecoregions." "We are looking at how to restore the environment and bring nature back into people's lives," he says.

A MORE RECENT LESSON FROM CONSERVATION HISTORY

The feeling of "saving ourselves" also had been expressed by Steve Young in describing the battle later in the 1970s to save the nation's last great wildlands in Alaska. Ten years after he coordinated Alaska Coalition grassroots efforts, Steve shared his recollections about the legislative battle with *Earth Work* while sharing tea and chili on an autumn afternoon in a Vermont barn. Young lives far from the lobbying hubbub of Washington, D.C., now, preferring his country home and his current job as a regional representative for the National Audubon Society. Nonetheless, a discussion about the concerted battle by virtually all national conservation organizations to pass the Alaska National Interest Lands Conservation Act of 1980 (ANILCA) transported him back in time. He recalled the fervor of working for nonprofit organizations in the 1970s: "There was a romance to it, and there was the magnitude of saving our last frontier. To people in the lower forty-eight states, it was as if a hundred years ago we had saved the area west of the Rockies, as if we were saving ourselves."

In fact, the battle to save Alaska's national parks, wildlife refuges, and wilderness in the late 1970s was unprecedented in several ways. First, there was the sheer size of the endeavor: The historic Alaska Lands Act set aside 43.6 million acres for the National Park System and 53.7 million acres for the National Wildlife Refuge System. It also created an overlay of 56.6 million acres of wilderness within the conservation units and designated twenty-six wild and scenic rivers.

Second, there was the symbolism: our nation's last great wild expanse, our final chance to preserve whole ecosystems, to do it right the first time. Third, there was the Alaska Coalition itself: a band of conservationists eager to take on America's most powerful extractive industries—big oil, mining, and timber and their allies in Congress and the state of Alaska.

Though largely young, this band waged an environmental campaign unprecedented in scale or sophistication. Lessons learned in the Alaska Coalition period have a strong influence on how conservation organizations conduct business today. The general agreement is that the conservation movement has incorporated lobbying techniques developed during the Alaska campaign into other campaigns and learned to do some things better, but that the Alaska Coalition is still the best example of an integrated, grass-roots success. At hearings in the lower forty-eight states and Alaska, the coalition attracted many of the more than a thousand witnesses and reportedly involved more citizens than any effort since the civil rights movement of the 1960s.

The Alaska campaign gave conservationists such as Steve Young a firm belief in the grassroots and attracted people who had never before considered themselves conservationists.

Sitting in her large, bright, plant-filled office at the National Wildlife Federation headquarters in Washington, D.C., Sharon Newsome recalled how twenty years ago she was "someone who never cared about the environment." She was, however, intent on living off the land, so she and her husband leased some land from the Forest Service, built a cabin with their own hands, and went hunting and subsistence fishing along a beautiful river running through the cathedral rain forest in the Tongass. Upon discovering that the Forest Service had plans to log the area, Newsome "became an environmentalist. I became involved and sophisticated." She went to Washington for one of the first Alaska Coalition training colloquia, testified at hearings in Ketchikan—where bumper stickers read, "Sierra Club, kiss my axe"—and became president of the Tongass Conservation Society even though previous leaders had become outcasts. Newsome since has fought many conservation battles in the lower forty-eight. Today she is vice president of the nation's largest conservation organization, but a conservation career was not in her original plans.

Chuck Clusen, co-chairman of the Alaska Coalition during the 1970s legislative battle, and Destry Jarvis, Senate lobbying coordinator and then coalition chairman during the early 1980s, both planned conservation careers, but the Alaska battle was pivotal in their rise to their current leadership positions. Both have been working on Alaskan conservation issues—among others—for most of their careers. Clusen's interest began with lobbying on the Alaska Native Claims Settlement Act of 1971 for the Sierra

George Matsumoto, Monterey Bay Aquarium Research Institute
Earth Worker

George I. Matsumoto, Ph.D, 31, is a volunteer interpreter for the aquarium and a visiting scientist at the Monterey Bay Aquarium Research Institute (MBARI). He began his career as a high school volunteer with the Student Conservation Association at Yosemite National Park in California and then as a paid SCA supervisor at Golden Gate National Recreation Area. Today he works with students of all ages as he interprets the images in the "Live from the Deep Canyon" exhibit at the Monterey Bay Aquarium in California. Through his interpretation and his scientific job, George is directly involved with both the exhibit and the research.

The exhibit "Live from the Deep" is the first permanent one of its kind, made possible by the innovative technology and resources of the Monterey Bay Aquarium Research Institute (MBARI). Through a microwave linkup between MBARI's underwater Remotely Operated Vehicle, a computer interface, and a laser disc that

continued

Club and continues in his present job as a senior associate at the Natural Resources Defense Council and a member of the board of directors of the American Conservation Association.

Jarvis, former executive vice president of the Student Conservation Association and former vice president of the National Parks and Conservation Association, has found his nonprofit policy-making experience invaluable in his new position as special assistant for policy to the director of the National Park Service. He points out that the Alaska Coalition was the first community-wide sophisticated lobbying effort. In a series of meetings in 1975–76, conservationists in Alaska and in the nation's capital decided that their only chance was an unprecedented national mobilization. They got agreement from the CEOs of their organizations to contribute financial resources and lend full-time and part-time staff power to a group effort on a new scale. Up to that point, conservation organizations had functioned independently of one another; when they did come together, it was to co-sign joint-issue letters. There had never been a formal coalition, in the sense of individual organizations submerging their identities for the sake of united action, on an issue for such a long time.

"For four years," Jarvis underscores, "the separate organizational identity, organizational competition, and personal rivalries that inevitably exist were submerged or secondary to the cause of enacting the Alaska Lands Act. That unanimity gave the Alaska Coalition unprecedented visibility, credibility, and respect on Capitol Hill, in the media, and among the public in general."

Clusen points out that working for nonprofits is different today. "Environmental groups are infinitely larger than they were ten years ago. They have far more technical expertise (economists, foresters, and so on), many more members, much larger budgets, and they also have bureaucracies. . . . There is a lot of good in institutionalization of environmental organizations and in new expertise, but institutionalization also gets in the way of being effective because it gets in the way of collaboration when groups are competing for money and media."

Clusen's main concern is that core conservation goals sometimes get lost because of a lack of collaboration and that organizations need to learn "that you can have it both ways—you can have institutionalization and also collaboration within and reaching without to labor and other groups."

continued from page 99

displays images, scientists and aquarium visitors are able to explore together the mysterious species in the Monterey Submarine Canyon.

George enjoys his interpretive work most of all. "You never know what you are going to see; everyone is excited. I really enjoy bringing this information to the public."

He received his bachelor's in marine biology from the University of California, Berkeley, in 1984 and his Ph.D in 1990 from the University of California, Los Angeles.

As a graduate student of marine biology at UCLA, George traveled extensively. Research expeditions involving scuba diving and blue-water diving took him around the world, from Catalina Island to Antarctica, to Iceland and Norway, and beyond.

Most rewarding to George of his many accomplishments have been his educational experiences, including the description of a new species. At Catalina, George discovered a new species in the phylum Ctenophora, a beautiful, fragile, jellylike animal not able to be collected by traditional means, which he named, described, and included in his dissertation. His work is widely recognized and has received

continued

When it was time for Steve Young to leave our meeting in a Vermont barn, he wanted to underscore why he came to Vermont instead of staying in Washington or at the headquarters of a conservation group: "I chose to come back to the grass roots because I like being a catalyst to show people that they can make a difference, that their dreams can happen, that they don't need to be an expert, that they can pick up the phone. That is the essence of the conservation movement. It begins in people's living rooms, not from the top down. We must never forget that."

continued from page 100

many honors and research grants.

George has also worked as part of a team teaching a novel deep-sea biology course at Hopkins Marine Station at Stanford University.

George encourages graduate study. And if you are interested in a particular professor's work or in working with a particular person, George suggests, "Don't be afraid to take the first step by writing, phoning, or somehow contacting people whose work you are interested in."

TURNING POINTS FOR NONPROFITS

President Jimmy Carter signed the Alaska National Interest Lands Conservation Act as a lame duck at the end of 1980. After their heyday in the 1970s, the policies and even the personnel of many nonprofit conservation groups came under attack by the Reagan administration during the 1980s. Some organizations adapted by becoming more conciliatory and using corporate techniques. For other organizations, marketing economics and negotiations with industries reflected a recognition of the need for sustainable development and new technologies. Still others became uncomfortable with what they considered overaccommodation of industry and called for a new return to the spiritual roots of the environmental movement, hearkening to Thoreau's call. They said fighting pollution and preserving lands necessitate conflict with purely economic concerns.

When Reagan's interior secretary James Watt attacked their conservation and environmental policies during the 1980s, the membership and staffs of nonprofit conservation organizations such as The Wilderness Society and the Sierra Club grew rapidly. Other notable developments during the decade included the founding of a new policy research center in 1982, the World Resources Institute, which provides well-respected reports to counteract rhetoric on all sides, and the growth of a new grassroots social justice movement concerned about the inordinate effects of toxic wastes and other pollutants on poor people and people of color. This movement has brought many new participants into environmentalism. The Citizens' Clearinghouse on Hazardous Wastes was formed in 1981 by Lois Gibbs, who first got involved as a housewife organizing people concerned about Love Canal. Today the Clearinghouse deals with more than a thousand groups concerned about hazardous wastes alone.

Earth Day 1990 bore further evidence that a new generation was entering the environmental movement in droves—but not necessarily through joining the well-known national groups. In just a few years, the Student Environmental Action Coalition has grown to include more than thirty thousand students from fifteen hundred campuses who organized themselves without the leadership of the national groups.

In 1992–93, many of those from the nonprofit community who had protested the environmental policies of the past twelve years were recruited by the new Clinton administration, creating vacancies at nonprofits and new dynamics between government and nonprofits. President Clinton and Vice President Gore promised radically different and pro-environment policies. Many of the administration's appointees at the Department of Interior, Environmental Protection Agency, Department of Agriculture, and other agencies had built their careers at nonprofits. (Others were young enough to be starting off with a fresh—and green—slate.) The CEOs of conservation and environmental groups—especially those known as "the Green Group"—became regular guests at the White House.

These events have necessitated many adjustments in the operations of nonprofits; the politics of dealing with a friendly administration are different from those of dealing with an administration openly opposed to most of the nonprofits' ideas. Politics are especially difficult when an organization disagrees with the "friendly" administration's policies and when nonprofits do not have unanimity about their own policies *or* those of the administration and may have lost touch with their own memberships.

The Conservation Fund has completed several noteworthy studies of nonprofit organizations, concluding that the larger organizations need more sophisticated systems for communicating with the grass roots: "Size can be a strength and a weakness. As groups grow, bureaucracies grow. But size does not correlate with efficiency. Local and grassroots organizations outside the [Washington, D.C.] beltway are making things happen. The reason for bureaucracy is not size but a lack of effective management."

The Conservation Leadership Project report repeatedly refers to the need for the conservation movement "to help convert the conservation ethic into a national ethos" and to "rediscover the

Tish Morris, New Mexico Museum of Natural History
Earth Worker

Looking back on how she got interested in environmental work, Letitia "Tish" Morris explains that she is a product of the environmental movement of the 1970s. "Our elders and the big businesses were messing up the world," she reminisces, "so I decided to go out and see if I could save the earth."

Upon receiving her B.A. in natural history from Prescott College in 1980, Tish participated in an SCA resource assistant program at Natural Bridges National Monument in Utah, where she did a little bit of everything. "It was a really intense and rich learning experience. Living and breathing that park day and night, I learned about the environment, geology, and archaeology of the area. I did everything from working at the visitor information desk, to interpretation, to going on patrol (hiking and on bike)," she says. Tish's SCA experience and later seasonal work with the Forest Service strengthened her desire "to educate the public about the world around them."

The combination of her environmental interests, people

continued

ethical foundations of [the movement's] mission." The question is, Will the newly savvy and financially competitive conservation organizations ever be able to cooperate again in a major legislative campaign against economic interests?

In 1991 at Golden Gate National Recreation Area in San Francisco, a statue of the late conservationist Rep. Phillip Burton (D-Calif.), author of many conservation laws, was dedicated. It is a monument not only to Burton but to his warning to nonprofit lobbyists in 1979, "The only way to deal with exploiters . . . is to terrorize the bastards." The second half of this quote is hidden in the pocket of the statue, leaving one to wonder if it will stand the test of time.

In 1992, Harvard professor Edward O. Wilson, author of the bestselling *The Diversity of Life,* said that, under what he calls a New Environmentalism, "except in pockets of ignorance and malice, there is no longer an ideological war between conservationists and developers. Both share the perception that health and prosperity decline in a deteriorating environment." Burton would have considered Wilson's "pockets" of ignorance more like deep canyons. How will nonprofits reconcile the advice of politicians and scientists? Will the battle lines ever be drawn in quite the same way in a world of "sustainable development"?

NONPROFIT ORGANIZATIONS GROW UP

The good news is that it is now respectable to work for a nonprofit organization. Gone are the days when family members would ask nonprofit employees, "When are you going to get a real job?" Reporters will no longer ask you if "acid rain" is the name of a rock band or if you care more about animals than people. Concern for the environment is mainstream. The American public is strongly behind controlling pollution, protecting wildlife, saving more of our land from development, and other actions once considered extreme.

continued from page 102

skills, and education evolved into Tish's present successful career as education specialist at the New Mexico Museum of Natural History in Albuquerque, where she has worked for the past eight years. Among her many duties at the museum, Tish trains volunteers to perform interpretation, directs field-oriented summer camps that promote learning science through exploring the natural environment, and teaches programs designed to assist teachers in presenting scientific information to their students. Tish says that she and the volunteers who assist her get as excited as the adults and children who visit this hands-on museum. "I enjoy working with the volunteers and dealing with the variety of people who visit the museum," says Tish.

Likening her present work to her SCA experience of twelve years ago, Tish says she is still doing what she enjoys and does very well. "Parks and [natural history] museums are alike—both do something to inform the public; we just happen to have exhibits," she says.

The good news for all of us is that this popularity and unprecedented political clout give nonprofit organizations more leeway to accomplish their goals. Mother Earth is no longer a heroine only in health food stores; she is fashionable in the White House and even on Wall Street.

The bad news is that over the past ten years, it may have become *too* respectable to be an environmentalist in some quarters. Some of the larger nonprofits are run like corporations—with the advantage of increasing professionalism and organizational development but the disadvantage of loss of ardor and the sense of purpose through which organizations are founded in the first place.

Conservation Fund president Patrick Noonan explained to *Earth Work* that this conflict means, "The conservation movement is beginning to reach maturity. The '70s and the '80s were a very emotional time. The environment is now a household word, and more and more people want to be involved. We now realize environmental values are a social cause. Our organizations have mushroomed, bringing the need for professional management. But idealism has given way to management too often. What we don't have is a blending of the ethos, the idealism, with managerial skills and talents." Leadership is needed to accomplish that blending. Noonan says, "To be a leader, the first thing is that you must passionately care about something, and if you passionately care about it, you want to make a difference, and if you want to make a difference, you engage in the cause, and if you engage in the cause, you are going to understand it, its needs and networks, and how to make things happen."

How does one get started at such a profound task? He says, "When I was a young person, the stream I used to fish in Montgomery County, Maryland, got polluted. I cared about it, and I couldn't understand why this was allowed to happen. I read about The Nature Conservancy in the newspaper and called them up. I latched onto it. You latch onto it and see how it networks, who the players are, what makes it happen. I passionately cared. I found a mentor. I moved up. My message to a young person or a young organization is the same: You latch onto a big whale."

No generation has a monopoly on idealism or organization. Today it is evident on college campuses and in the offices of both large and small nonprofits across the nation that nonprof-

Tina Berger, Sport Fishing Institute
Earth Worker

Marine biologist Tina Berger, 29, is a research specialist and director of the Sport Fishing Institute's artificial reef development center. Her diverse jobs make her a key player on SFI's fifteen-person team, which works to protect the interests of sport fishermen through fisheries and habitat conservation and education. Berger's job requires that she travel to meetings and conferences to work on legislative issues and to meet with state fisheries managers and interstate commissions, a schedule that regretfully precludes her from actually fishing.

She did her undergraduate work in biology with a marine biology concentration at Boston University's Marine Program, which gave her the opportunity to work and study at the Marine Biological Laboratory (MBL) in Woods Hole, Massachusetts, for a semester. There, she completed concentrated studies with four professors, including fishing for bottom-dwelling creatures. This is where she got to spend "intensive time with different organisms," she joked.

continued

its will adapt, survive, and prosper as a sign of the American people's growing respect for all the species who make up the organism we call Earth.

CONSERVATION LEADERSHIP PROJECT

Based on interviews with more than five hundred CEOs and other leaders of conservation groups, The Conservation Fund made the following findings:

continued from page 104

Berger went to the University of Rhode Island to get her M.S. in marine policy, taking classes in fisheries management, ocean law, oceanography, and marine geology. While working on her thesis, she did a six-month internship at the National Wildlife Federation, in Washington, D.C., to work on marine fisheries, conservation, and wetlands issues, especially concerning tuna and shark. She finished her thesis, "Perceptions and Behaviors in New York's Lobster Fishery, a Social Study of Fishery-User Groups," and got her M.S. in 1990. She began work at SFI the same year.

- ✦ Many of the significant ideas and effective techniques for advancing environmental policies and programs have emerged from the grassroots levels of nongovernmental and nonprofit organizations.

- ✦ Too few membership organizations effectively activate their members, while nonmembership organizations give little thought to accountability to the grass roots, "confusing expertise with accountable public service."

- ✦ "Virtually none of the mainstream groups . . . works effectively with, or tries to include, people of color, the rural poor, and the disenfranchised."

- ✦ The typical conservation CEO comes to the job with no formal preparation and "matriculates at the trial-by-error school of management." Nearly all training opportunities in academia are designed for midcareer professionals in government, consulting firms, and corporate environmental programs. Training for nongovernmental organization (NGO) conservation careers is an afterthought.

- ✦ Typical leaders queried for the Conservation Leadership Project study spend less than 10 percent of their time on public speaking, media relations, and other outreach activities to nurture an environmental ethos. They spend little time on long-term planning and organizational development.

The Conservation Leadership Project suggests the simultaneous pursuit of six distinct strategies:

1. *Provide effective in-service training and leadership development to current leaders of the movement.*

2. *Attract new leaders with scholarships for good students, internships, and recruitment from other fields.*

3. *Create new academic programs for training leaders.*

4. *Revitalize volunteerism at all levels.*

5. *Expand participation by minorities and other nontraditional groups.*

6. *Work creatively with churches, schools, businesses, media, and other sectors to institutionalize the environmental perspective, making it central to the national ethos and a key constituent of the American character. Revitalize student conservation organizations across the nation.*

A DIRECTORY OF SELECTED NONPROFITS

Air and Waste Management Association, 3 Gateway Center, Four West, Pittsburgh, PA 15230, (412) 232-3450

American Association of Zoological Parks and Aquariums, Oglebay Park, Wheeling, WV 26003, (304) 242-2160

American Farmland Trust, 1920 N Street NW, Suite 400, Washington, DC 20036, (202) 659-5170

American Fisheries Society, 5410 Grosvenor Lane, Suite 110, Bethesda, MD 20814, (301) 897-8616

American Forests (formerly American Forestry Association), 1516 P Street NW, Washington, DC 20005, (202) 667-3300

American Rivers (formerly American Rivers Conservation Council), 801 Pennsylvania Avenue SE, Suite 400, Washington, DC 20003-2167, (202) 547-6900

Audubon Naturalist Society of the Central Atlantic States, Inc., 8940 Jones Mill Road, Chevy Chase, MD 20815, (301) 652-9188

Carrying Capacity Network, Inc., 1325 G Street NW, Suite 1003, Washington, DC 20005-3104, (202) 879-3044

Center for Marine Conservation, Inc., 1725 DeSales Street NW, Suite 500, Washington, DC 20036, (202) 429-5609

Chesapeake Bay Foundation, Inc., 162 Prince George Street, Annapolis, MD 21401, (410) 268-8816; Heritage Bldg., Suite 815, 1001 E. Main Street, Richmond, VA 23219, (804 780-1392); 214 State Street, Harrisburg, PA 17101, (717) 243-5550; 100 W. Plume Center #701, Norfolk, VA 23510, (804) 622-1964

Jeff Olson, The Wilderness Society
Earth Worker

Jeff Olson is representative of a new arm of the environmental community: the professional resource economist. He directs the Arnold Bolle Center for Sustainable Forest Ecosystem Management of The Wilderness Society, managing a team of seven ecologists and economists engaged in applied policy research.

Before joining TWS, he planned timber sales for a timber company and worked on forestry issues for the state of Michigan. At the Society he has been especially active in the top domestic forestry issue of the day: the fight to save the ancient forests. He has been involved in the spotted owl court cases.

Jeff spent two years in Oregon working on a study of the transition now occurring in the Northwest timber industry and how local economies can adjust to new realities. As a result, he has generated highly respected estimates to counter the job-loss claims issued by the timber industry during the debate over forest protection. Ben Beach, TWS deputy director of public affairs, says Jeff makes his job

continued

Clean Water Action, 1320 18th Street NW, Washington, DC 20036, (202) 457-1286

Conservation Foundation, The, 1250 24th Street NW, Washington, DC 20037, (202) 293-4800

Conservation Fund, The, 1800 North Kent Street, Suite 1120, Arlington, VA 22209, (703) 525-6300

Conservation International, 1015 18th Street NW, Suite 1000, Washington, DC 20036, (202) 429-5660

Cousteau Society, Inc., The, Headquarters: 870 Greenbrier Circle, Suite 402, Chesapeake, VA 23320, (804) 523-9335

Defenders of Wildlife, 1101 14th Street NW, Suite 1400, Washington, DC 20005, (202) 682-9400

Earth Island Institute, 300 Broadway, Suite 28, San Francisco, CA 94133, (415) 788-3666

Environmental Action Foundation, Inc., 1525 New Hampshire Avenue, Washington, DC 20036, (202) 745-4870

Environment Action, Inc., 6930 Carroll Avenue, Takoma Park, MD 20912, (301) 891-1100

Environmental and Energy Study Institute (EESI), 122 C Street NW, Washington, DC 20001, (202) 628-1400

Environmental Careers Organization, Inc., The (formerly CEIP Fund), 286 Congress Street, 3d floor, Boston, MA 02210, (617) 426-4375

Environmental Defense Fund, Inc, Headquarters: 257 Park Avenue South, New York, NY 10010, (212) 505-2100

Environmental Law Institute, The, 1616 P Street NW, Suite 200, Washington, DC 20036, (202) 328-5150

Greenpeace USA, Inc., 1436 U Street NW, Washington, DC 20009, (202) 462-1177

Humane Society of the United States, The, 2100 L Street NW, Washington, DC 20037, (202) 452-1100

continued from page 106

easier: "With people like Jeff on our side, we can make the hardheaded case that what is best for the environment is generally best for the economy."

Jeff says he left the timber industry to join TWS as a resource economist in 1987 because he was "at a time in my career when I was not really happy with the approach of industrial forestry." Having worked where the industry was "killing as many hardwood trees as found and converting the land to loblolly plantations," he concluded that was not what he became a forester to do. "Most people go into forestry because of a real concern for the environment. The job of industrial forestry is timber farming and is not related to protecting the environment. But The Wilderness Society gave me an opportunity to become involved in policy deliberations that affect the way timber harvesting is done."

International Institute for Energy Conservation, 750 First Street NE, Washington, DC 20002, (202) 842-3388

Izaak Walton League of America, Inc., The, 1401 Wilson Blvd., Level B, Arlington, VA 22209, (703) 528-1818

League of Conservation Voters, 1707 L Street NW, Suite 550, Washington, DC 20036, (202) 785-8683

Mineral Policy Center, 1325 Massachusetts Avenue NW, Suite 550, Washington, DC 20005, (202) 737-1872

National Parks and Conservation Association, 1015 31st Street NW, Washington, DC 20007-4406, (202) 944-8530

National Recreation and Park Association, 2775 S. Quincy Street, Suite 300, Arlington, VA 22206

National Wildlife Federation, 1400 Sixteenth Street NW, Washington, DC 20036-2266, (202) 797-6800

Natural Resources Defense Council, Inc., 40 West 20th Street, New York, NY 10011, (212) 727-2700

Nature Conservancy, The, 1815 North Lynn Street, Arlington, VA 22209, (703) 841-5300

Rails-to-Trails Conservancy, 1400 Sixteenth Street NW, Suite 300, Washington, DC 20036, (202) 797-5400

Rainforest Action Network, 450 Sansome, Suite 700, San Francisco, CA 94111, (415) 398-4404

Renew America, 1400 Sixteenth Street NW, Suite 710, Washington, DC 20036, (202) 232-2252

Resources for the Future, 1616 P Street NW, Washington, DC 20036, (202) 328-5000

Sierra Club, 730 Polk Street, San Francisco, CA 94109, (415) 776-2211

Student Conservation Association, Inc., National Headquarters: P.O. Box 550, Charlestown, NH 03603, (603) 543-1700

U.S. Public Interest Research Group, 215 Pennsylvania Avenue SE, Washington, DC 20003, (202) 546-9707

World Environment Center, 419 Park Avenue South, 18th floor, New York, NY 10016, (212) 683-4700

World Resources Institute, 1709 New York Avenue NW, Washington, DC 20006, (202) 638-6300

World Wildlife Fund, 1250 24th Street NW, Washington, DC 20037, (202) 293-4800

Abby Arnold, RESOLVE
Earth Worker

Abby Arnold is an SCA volunteer who worked in Merck Forest in Vermont in the High School Program some five years after Bruce Hamilton. A resident of Los Angeles, she saw it as a way to "get into the wilderness."

Today she is an environmental mediator in the RESOLVE organization, an independent program of the World Wildlife Fund. She emphasizes that she is a neutral facilitator, siding with neither party in disputes over issues such as facility siting, public policy on biofuels, and access to fisheries. She was well prepared for her unusual role by previous experience at Conflict Management, Inc., in addition to earning a master's in public administration from Harvard (1988) and a B.A. in environmental studies from the University of California at Santa Cruz. At a recent International Union for the Conservation of Nature (IUCN) Parks Congress in Caracas, Venezuela, she led negotiation training workshops.

Chapter Seven

JOBS IN ENVIRONMENTAL MANAGEMENT

Who are today's earth workers? For most of us, the phrase conjures up images of field scientists working to conserve the plants and animals of our great wilderness areas, forests, national parks, and oceans. Activists and lawyers at our national conservation groups would be part of the picture. Outdoor recreation leaders, educators, and naturalists would be included, of course. We might round out the picture with the hundreds of employees at conservation districts, land trusts, and conservancies who help protect our local pieces of the earth.

Beyond these land and water conservation professions, however, there is another group of environmental specialists devoted to reducing the use of toxic materials, cleaning up hazardous waste sites, preventing air pollution, dealing with solid waste, removing asbestos and lead from our buildings, and promoting recycling. This is a growing part of the environmental job market.

LEGISLATION: THE ENVIRONMENTAL JOB GENERATOR

Since the passage of the National Environmental Policy Act in 1970, and the creation of the U.S. Environmental Protection Agency, our national, state, and local legislatures have passed a dizzying array of environmental laws and regulations affecting nearly every area of our lives. The Clean Air Act, Clean Water Act, Resource Conservation and Recovery Act, Superfund Amendments and Reauthorization

Act, and others have launched an army of new environmental professionals into industry, government, consulting firms, and nonprofit groups.

These environmental careers are among the fastest growing professions for the 1990s and beyond. What kind of people are needed? Where do they work? How can you get started as an environmental professional? In this chapter, we will take a look at four areas with a profile of a current professional in a given field. Starting with solid waste management, we will move on to air quality, water quality, and toxic material reduction and hazardous waste management.

SOLID WASTE MANAGEMENT AND RECYCLING

One statistic tells volumes about our solid waste crisis: Fully half of the nation's fifteen thousand landfills are expected to be filled and closed by 1995. Municipalities throughout the country are seeking alternatives to the local dump and finding them in waste reduction, recycling, "waste-to-energy" plants, incinerators, and others. Left behind are hundreds of landfills, many of which are severely contaminated and require cleanup.

The solid waste problem is creating new jobs at a steady pace. A 1992 edition of *U.S. News and World Report* listed "recycling coordinator" as one of the fastest growing careers, and although the pace has slowed, recycling progress is still strong. Also needed are engineers of all kinds, policy managers, transportation analysts, chemists, and an army of traditional "garbage people" to pick up, transport, and sort out our trash. Finally, large numbers of environmental attorneys will be kept busy alternately seeking out and fighting against sites for new incinerators and landfills.

A CAREER AS A RECYCLING COORDINATOR

How do you get started? Conversations with current recycling coordinators show that workers come from a wide variety of different backgrounds, including general liberal arts, geography,

Carol O'Dahl, Business and Industry Venture
Earth Worker

Carol O'Dahl was thinking diapers, oil spills, Kuwait, and the man with the eye patch. She had been staffing a booth at Seattle's first "Buy Recycle" Trade Show when a tall and extremely frustrated person asked for her help. His company, it seemed, made oil spill retention booms out of recycled baby diapers, and he was having a hell of a time getting them to the oil-soaked waters of the Persian Gulf.

No problem. Carol cut through the red tape and got in touch with Boeing for help and transport, and soon the former diapers were soaking up crude in the Middle East.

Why ask Carol? Now in her sixth year as a recycling professional, she is one of two "recycling information specialists" for the innovative Business and Industry Recycling Venture, a public-private partnership sponsored by the Greater Seattle Chamber of Commerce in coordination with city and county solid waste agencies.

"My job is to provide information and referrals to King County business to help them recycle their own waste and buy products made from

continued

planning, environmental studies/science, business/economics, and various technical fields. As the field develops, we can expect "professional" standards, academic programs, two-year degrees, and other rites of passage to spring up. A few key requirements are already becoming evident.

Recycling programs focus on several areas: program planning and design; creating public awareness; logistics of pickup, sorting, and storage; finding or creating markets for recycled materials; transportation; and overall management. Each one requires superb organizational skill, oral and written communication ability, the ability to work skillfully with a wide range of people, computer literacy, and mastery of the skill unique to your subarea.

Local and regional recycling programs report a particular demand for coordinators who can assist in market identification or creation. As more and more material is collected for recycling, current markets are saturated, reducing prices for recycled materials or eliminating markets altogether. In the worst case, materials collected for recycling end up right back in the landfill. Creative municipalities are hiring people who help businesses and government agencies purchase finished items made from recycled materials. This creates a need for manufacturers to invest in production using recycled raw material and thus raises demand for the paper, glass, plastic, and metal collected by local recycling programs.

The cycle of job creation from recycling keeps expanding—from the manufacturers making new products from old materials to the brokers who seek out and purchase the raw materials, to the cities and towns (or their contractors) that collect and store recyclables, to the journalists, teachers, marketing people, and advocates who convince us to recycle to the consumer, all the way to the voters who make the whole thing go.

HOW TO GET STARTED

If a career in recycling or solid waste management sounds like a good idea, the easiest way to get started is to get some experience as a volunteer or intern. Look into your local situation. The chances are you will uncover a collection of private businesses, consultants, government agencies, legislative committees, nonprofit groups, and neighborhood associations that would welcome your involvement.

continued from page 110

recycled material. We do consultations with businesses, publish recycling guides and product lists, sponsor the trade show, answer questions—whatever it takes," says Carol. "The response has been overwhelming. Businesses like Nordstrom, Eddie Bauer, Seattle First National Bank, and others really want to get involved."

Carol's career path has been a fascinating one. After graduating from the University of California at Santa Barbara in 1985 with a B.A. in environmental studies, she went to work as a canvasser for the League of Conservation Voters, a job that she says "really made a difference. In my current job," she explains, "being organized, managing people, motivating others, thinking on your feet, and most importantly, communicating well are crucial. Canvassing helps you develop all of those skills."

From there, Carol had two recycling-related project positions with city and county government in Seattle through the Environmental Careers Organization. Those positions gave her the experience she needed to understand local government, recycling issues, and how to get things done. When she finished her second

continued

To keep up on the rapidly changing world of recycling, one of the best resources is the bimonthly *Resource Recycling* magazine, P.O. Box 10540, Portland, OR 97210.

continued from page 111

project, she had the knowledge and confidence she needed to approach the University of Washington about her newest idea: a comprehensive recycling plan.

For the next two years, Carol was on the staff of the Pacific Energy Institute, a nonprofit environmental organization in Seattle. Among other things, she researched and authored a home waste guide. When Business and Industry Venture opened its doors, she was a natural candidate for the job. Carol's advice to others? "Don't be afraid to take risks."

WATER QUALITY

The images are indelible: dead fish floating in a sea of greasy oil and foaming phosphates, untreated waste shooting from a rusted pipe into a polluted lake, burning rivers, "No swimming or fishing" signs. Images such as these from the late 1960s went a long way toward creating the infrastructure of laws, regulations, and agencies that now have had more than two decades to improve the quality of our water resources.

What are the water issues of the 1990s? What kind of career opportunities are available? How can you prepare yourself for a job in the water quality field? To answer some of these questions, we talked to current employers and professionals in government, industry, and the nonprofit world. We found a promising future for people who care about clean, plentiful water.

One useful way to think about water quality careers is to view the organizational chart for the U.S. Environmental Protection Agency's water-related programs. EPA has set up separate offices for (among others)

- ↝ Drinking water
- ↝ Groundwater protection
- ↝ Marine and estuarine protection
- ↝ Municipal pollution control
- ↝ Wetlands protection

Within pollution control, we can make an important distinction between "point" and "non-point" source control. In certain parts of the country, we can also add serious concerns about water quantity, which will not be addressed here.

MUNICIPAL POLLUTION CONTROL

Until the early 1980s, most water quality efforts were focused on building and monitoring municipal and industrial waste treatment facilities to reduce emissions of "conventional" pollutants into lakes, rivers, and oceans. Progress on this front has been remarkable and is one of the great success stories of the environmental movement. It has also created a need for thousands of trained managers and technicians to run the treatment plants, collect and tèst samples, and ensure compliance with federal and state regulations.

In many ways, these professionals are unsung heroes in our common struggle for a cleaner environment. After the lobbying is done and the legislation is passed and the funding is allocated and the "policy options" are "analyzed" and the grant is made and the construction is completed, it is the wastewater technicians and lab chemists who do the real day-to-day work of ensuring cleaner water. If you have technical aptitude and are looking for relatively quick entry into the environmental field, an associate's degree in wastewater technology from a local community college may be for you.

TOXICS CONTROL AND GROUNDWATER PROTECTION

Although great progress has been made in controlling the emission of "conventional" pollutants, we have just begun to understand and reduce the release of a large number of toxic pollutants whose overall effect on water quality may be much greater. Since the passage of the Clean Water Act amendments of 1987, the number of toxicants whose emission into our waterways is regulated by the government has grown substantially. Continued research undoubtedly will identify more.

The focus on waterborne toxicants is generating a new group of environmental professionals in such fields as toxicology, risk assessment, environmental chemistry, fisheries science, environmental engineering, and regulatory analysis. According to the human resources director of a major consulting firm, people with advanced degrees in these fields are in high demand. "Risk assessment specialists are particularly hard for us to find," she says. "We need to gain a better understanding of what risks are represented by trace quantities of toxicants in our water sources."

A field that is growing even more rapidly is groundwater analysis and remediation. Consider the following facts and the big picture begins to come into view:

Sam Pett, Tetratech
Earth Worker

Sam Pett grew up in Michigan, where he enjoyed swimming, canoeing, and sailing on that state's many inland lakes. Today the 35-year-old biologist is combining his interest in marine biology and his love for the water. As an employee at the Fairfax, Virginia, office of Tetratech, a national environmental consulting firm, Pett reviews and presents scientific data to assist the EPA in meeting the policy objectives of various federal laws, such as the Ocean Dumping Act and the Clean Water Act.

Pett's route to his current job was decidedly circuitous. After graduating from Michigan State University in 1978 with a B.S. degree in wildlife biology and zoology, Pett did a two-year stint with the Peace Corps in Niger, West Africa, to work on agricultural pest control and parks management. He then took a job with a housing trust in Boston to create permanent housing for inner-city homeless people. "But my heart was in biology and the environment, so I went back to school," he says.

At the age of 31, Pett entered the environmental

continued

↝ Nearly half of all people in the United States depend upon groundwater sources for their drinking water.

↝ There are more than one hundred thousand landfills in the nation, many of which contain toxic materials that may seep directly into groundwater aquifers.

↝ Up to 10 million underground storage tanks are "out there," and forty-one states have reported groundwater contamination from such tanks.

↝ About 700 million pounds of pesticides are used every year in America, and fully half the states have identified contamination from these chemicals.

Of the many "hot" disciplines working in the groundwater arena, none is more in demand than hydrogeology. Although chemists, hydrologists, engineers, computer scientists, and mathematicians (the latter simulate pollution plumes in groundwater) are all needed in dramatic numbers, hydrogeologists with advanced degrees are a particularly scarce breed.

Terry Hove, director of the California office of the Environmental Careers Organization, says, "Anyone with an M.S. in hydrogeology and some field experience can basically write his or her own ticket. We are discovering new groundwater problems every day and reassessing the severity of the ones we know about. It's an exciting field to be in and, not incidentally, a very lucrative one because of the shortage."

NON-POINT SOURCE CONTROL

When we think of water pollution, most of us think of brown pollution spewing out of an industrial pipe. Environmental protection means cleaning up the water that receives the pipe's discharge or, better yet, eliminating the polluted emission altogether. It is a comforting notion to think that we could solve our water problems by finding all the pipes and taking the owners to task.

Unfortunately, that is not even close to the truth. In 1984, the Conservation Foundation estimated that more than half the nation's water pollutants come from "non-point" sources that have only begun to be regulated. Determining how to stop the tons of heavy metals, nutrients, sediments, and chemicals that

continued from page 113

sciences program at the University of Massachusetts at Boston. During his studies, he went to work at EPA's Oceans and Coastal Protection Division as a Knauss Sea Grant fellow. There he reviewed project proposals from EPA's regional offices and produced reports for the National Estuary Program's office. His thesis analyzed the role of scientific research in managing sanctuaries through the National Oceanic and Atmospheric Administration's National Marine Sanctuary Program.

He went to Tetratech after his fellowship, at about the time he graduated. Pett says he was interested in private-sector work because he had always been in the public sector. During his three months in his new position as an environmental scientist, he has primarily worked on non-point source guidance for EPA. "EPA tells us where they need more information, then we work to get accurate data by collecting public comments, industrial studies, county and state studies, and journal articles."

Future opportunities with Tetratech might involve more project management responsibilities. "Eventually I'll probably end up working for a nonprofit again. I may be more suited for

continued

run off from agricultural fields, parking lots, streets, and other areas is a difficult, complicated task.

It requires a small army of new environmental professionals: "When we began this agency in 1987, we had seventy staff and a five-year plan to deal with the problem." The speaker is Ken Guy of the Surface Water Management Utility for King County in Washington State. "This year we will have two hundred people working for us, including ecologists, planners, biologists, geologists, engineers, geographic information systems (GIS) specialists, and others. The non-point side of the house is definitely growing."

Control of non-point pollution requires some cultural changes that can be brought about only through education. People with an interest in environmental education, community organizing, and like fields can be found organizing household hazardous waste days, painting storm drains with colorful "Do not dump!" signs, and using brochures and workshops to reduce the twenty-five gallons of hazardous chemical products used by each American household every year.

continued from page 114

that. I'm actually going to start volunteering soon for a committee working toward the creation of a national institute for the environment," he says.

Nevertheless, Pett believes that the paperwork he now spends his time on is part of a larger process to restore the nation's waters. "These guidance documents go out to all fifty states to help them develop non-point water pollution control programs. By reviewing the information and public comments and understanding the different perspectives, I can add something to the policy outcome that will benefit the marine and coastal environment."

NATURE'S WATER PURIFIERS: PROTECTING AMERICA'S WETLANDS

By now, it is common knowledge that we are losing our wetlands at an alarming rate. By some estimates, more than five hundred thousand acres of wetlands are drained, dredged, or paved over every year. The impact of this loss on aquatic life is obvious, but less appreciated is the effect of wetlands loss on water quality. Because wetlands serve as a natural water treatment process for many pollutants, their loss is a serious concern.

Wetlands protection is also a major growth area for environmental professionals. People with advanced degrees and field work experience in the biology, ecology, botany, and/or chemistry of wetlands are in demand. Similarly, cartographers, GIS specialists, and surveyors are needed to help define the always-controversial boundaries of wetlands areas. Finally, engineers with an interest in the design and construction of "artificial wetlands" are finding a growing demand for their services, as are environmental planners who can devise policies that find a compromise between the need to protect wetland resources and the demand for commercial and housing development.

FUTURE NEED

Water-related environmental careers are growing and will continue to grow for at least the rest of the decade. The need for qualified technical personnel, especially those with advanced degrees and field experience, is great. Those who combine scientific or engineering knowledge with the ability to manage, inspire, and communicate with people are needed most of all.

Good luck in your search. Clean water is the wellspring of all life. We need talented earth workers to ensure that it continues to flow.

RESOURCES

For more information on water-quality related careers, you may want to contact some of the following organizations.

American Water Resources Association, 5410 Grosvenor Lane, Suite 220, Bethesda, MD 20814

Association of State and Interstate Water Pollution Control, Administration, Hall of the States, 444 N. Capitol Street NW, Suite 330, Washington, DC 20001

American Water Works Association, 6666 W. Quincy Avenue, Denver, CO 80235

National Water Well Association, 6375 Riverside Drive, Dublin, OH 43017

Freshwater Foundation, 725 County Road 6, Wayzata, MN 44391

Cousteau Society, 930 W 21st Street, Norfolk, VA 23517

The Environmental Careers Organization, 286 Congress Street, 3d floor, Boston, MA 02210

AIR QUALITY

When more than thirty thousand environmental leaders gathered in Brazil for the United Nations Conference on Environment and Development (UNCED), air quality issues were among the most urgent discussed: global warming, the ozone hole, acid rain, airborne toxics. Air pollution concerns are in the news and have risen to the very top of our global environmental agenda.

These growing concerns are being reflected in a dramatic increase in job opportunities for environmental professionals. Formerly limited disciplines such as atmospheric science, air quality modeling, risk assessment, air chemistry, toxicology, industrial hygiene, air quality planning, and air-related engineering are now in hot demand.

With the passage of the 1990 Clean Air Act Amendments (CAAA), air quality careers got a big boost. The legislation set new directions for air pollution control and brought hundreds of new "sources," (i.e., businesses and vehicles) under increasingly stringent regulation. According to a recent article in *Environment Reporter*, CAAA will create sixty thousand new jobs in this decade.

Understanding the key provisions of CAAA will be a big help in determining the future of your air quality career. Each of CAAA's seven major areas will create new needs at private businesses, government agencies, and nonprofit organizations. Listed below are some of the crucial concerns covered in the regulatory titles.

IMPROVING AIR STANDARDS. More than twenty years have passed since Congress first set air quality standards for the most common pollutants, such as ozone, carbon monoxide, and particulate (the familiar cloud of smog that hangs over most U.S. cities). Yet two decades later, more than 100 million Americans live in cities that exceed important air pollutant limits. The CAAA tries to rectify this situation by imposing stricter standards and setting up new timetables for meeting them.

Don't look for a letup in employment generation anytime soon, however. Los Angeles and other "extreme" cases have twenty years to meet the new goals, and past history suggests that extension requests are almost inevitable.

REDUCING VEHICLE EMISSIONS. Cars, trucks, and buses account for a huge share of urban pollution, including 90 percent of carbon monoxide emissions. The CAAA puts our nation on a rigorous schedule for reducing pollution from "mobile sources." Many local governments are creating innovative plans to deal with the situation even more forthrightly, by getting cars off the road through development of alternative transportation.

CONTROLLING AIR TOXICS. Identifying, measuring, and controlling air toxics is a major new air quality field, and CAAA radically extends the range of emissions and sources being regulated. With nearly two hundred air toxicants having been identified for control so far, polluters in need of help range from huge oil refineries and incinerators to your local dry cleaning shop.

PREVENTING ACID RAIN. The U.S. Environmental Protection Agency (EPA) estimates that electric utility companies, one of the major employers of air quality professionals, produce 20 million tons of sulfur dioxide each year. Sulfur dioxide from utilities and other sources combines with atmospheric deposition and returns to us in the form of acid rain, harming lakes, streams, and ponds; reducing forest growth; and damaging crops. Under the CAAA, more than two thousand utility companies will be required to cut emissions roughly in half by the year 2000. Smelters and other "pipestack" industries will also be affected in a big way.

CREATING INCENTIVES. One of the most important parts of the new clean air amendments is Title 5, which requires all polluters to pay for permits based on their emissions. This innovation was designed to create an economic incentive to reduce pollution and to provide a valuable source of revenue for state and local air pollution programs. Title 5 is creating previously unheard-of jobs for economists, M.B.A.s, accountants, and other business types in private industry and at government agencies.

REDUCING OZONE DEPLETION. Although the political and scientific consensus on the severity of ozone depletion and the existence of "the greenhouse effect" and "global warming" is less than perfect, everyone agrees that we must reduce the emission of ozone-depleting compounds such as CFCs (chlorofluorocarbons), halon, and carbon tetrachloride.

The law requires that CFCs and HCFCs be phased out by the end of 1995, and other culprits by the year 2000. Career opportunities exist not only in eliminating these emissions but in finding alternatives to their commercial and consumer uses.

INCREASING ENFORCEMENT. The CAAA increases the severity of fines and other penalties for air pollution lawbreakers. Willful violation, for instance, is now a felony, and the possibility of jail time for executives is a real one. Would-be lawyers take note—a substantial increase in legal activity both in and out of court should be the result.

SKILLS IN DEMAND

By far the greatest growth in jobs in the air quality field will be in scientific, technical, and engineering positions. Technicians, for instance, are in great demand. These people, often trained at two-year colleges, carry out the essential work of gathering air data and preparing it for analysis, often performing chemical tests.

Risk assessment specialists and toxicologists are finding ample opportunities for their skills, especially in the air toxics arena. Determining the toxicity of various pollutants in the air, understanding their properties, assessing the risks to human health of different concentrations, and reducing their emissions will require sophisticated abilities, usually at the M.S. or Ph.D. levels.

Environmental engineers with air quality backgrounds are needed at all levels, as are process engineers who can assist us in developing pollution prevention approaches to our air quality problems.

The field will also need air quality planners who combine technical knowledge with managerial, legal, economic, and communication skills to develop comprehensive programs that attack air concerns along a broad front through synergistic actions.

Finally, as in most other environmental fields, analytical chemists will find a ready market for their degrees.

GETTING STARTED

Here are a few tips for identifying employers, finding out their needs, and preparing yourself for an air quality career.

↪ Understand the CAAA and its requirements.

↪ Identify local and state air quality agencies in your area. Most major cities have air pollution control agencies, many of which are set up as regional authorities.

Two or three well-planned informational interviews should garner you valuable information about their programs, hiring projections, and skill needs. Don't forget to ask about potentially nontechnical areas such as contract management, community relations, and policy analysis.

↜ Identify key polluters. Information from EPA, local air agencies, and newspaper stories can be used to help you key in on businesses that will be in need of help. Government staff people often are willing to share some company names with you. Visit environmental affairs departments at selected firms to get the lowdown on their efforts to reduce air pollution. Ask about their permit purchase programs, which may very well need help.

↜ Seek out consulting firms. Not surprisingly, a large number of small and large consultants are beefing up their air quality staffs in an effort to cash in on CAAA requirements. A simple place to begin is the local Yellow Pages.

NOW IS THE TIME

It is almost impossible to understate the consequences of inaction on our air pollution problems. As public awareness grows, the demand for meaningful action on local, national, and global air problems will increase, requiring a pool of trained professionals to address the issue. Preparing now will not only guarantee you a promising career but will give the rest of us greater hope for the future.

RESOURCES

The following organizations may be helpful to contact as you develop your air quality career.

Air and Waste Management Association, Box 2861, Pittsburgh, PA 15230

Association of Local Air Pollution Control Officials, 444 N. Capitol Street NW, Suite 307, Washington, DC 20001

California Air Resources Board, Box 2815, Sacramento, CA 95812

In addition, the Environmental Careers Organization (ECO) offers paid short-term positions for students and recent graduates in air quality as well as other fields. Write to 286 Congress Street, Boston, MA 02210.

HAZARDOUS WASTE MANAGEMENT

GROWING AN INDUSTRY AND SOLVING A PROBLEM

Twenty years ago, when many of today's college seniors were still in diapers, the field of hazardous waste management did not exist. Total expenditures on hazardous waste site remediation and restoration, regulatory compliance, and other staples of today's environmental budgets were near zero. There was no Superfund. No agency regulated the manufacture, transportation, or disposal of hazardous waste. The U.S. military had no idea that cleaning up its empire of military bases and weapons production facilities would cost tens of billions of dollars.

Today, there are no illusions. Everyone knows about hazardous waste. Each year, the nation's industry and military generate and dispose of over 400 billion pounds of hazardous chemicals, most of it carefully monitored and regulated. The images of poisoned land and water, of fields full of toxic drums, of uprooted towns like those around Love Canal and Times Beach have become part of our collective consciousness. Over 50 million Americans live within four miles of a Superfund hazardous waste site.

To deal with the problem, a small army of hazardous waste management professionals and laborers has grown up with incredible speed, fueled by the expenditure of billions of dollars in public and private funds. In 1994, the "market" for hazardous waste management and cleanup of old sites is likely to exceed $17 billion. The Department of Energy alone requested $6.5 billion from Congress in 1994 for environmental restoration projects, the majority of which will go to hazardous waste management and remediation activities.

Who are the "earth workers" in the hazardous waste profession? How many are needed? What do they do? Where do they work? What kind of academic preparation is needed to compete in this growing field? What trends are guiding the field in the mid-1990s?

TODAY'S HAZARDOUS WASTE FIELD

The numbers are telling. It is estimated that over 160,000 people will be directly employed in hazardous waste management in 1995, up from 93,000 in 1991. Working at government agencies, environmental consulting and engineering firms, nonprofit organizations, and manufacturing corporations, hazardous waste managers, scientists, technicians, advocates, and communicators are everywhere. The hazardous waste management profession, nonexistent two decades ago, is now a permanent part of the environmental career landscape.

The field can be roughly divided into two groups. People who manage the current flow of hazardous materials and waste make up one group. The other comprises people who are cleaning up current toxic messes and spills and remediating the toxic waste dumps of the past. Surprisingly, the high-profile activity of site remediation, which receives most of the attention in the popular press (and a great deal of federal, state, and private industry cash), involves fewer people than ongoing work of managing today's regulated waste.

DAY-TO-DAY HAZARDOUS WASTE MANAGEMENT

This ongoing work can become highly complex. Conceptually, however, it is quite straightforward. The key elements of successful "hazmat" programs involve identifying hazardous wastes used in production processes or present for other reasons; securing government permissions to produce, transport, or store identified hazardous materials; tracking hazardous materials; disposing of wastes according to relevant laws and guidelines; monitoring the dumps and other disposal sites to which wastes are sent; and communicating with community members and environmental advocates to maximize public involvement.

This "cradle-to-grave" approach to hazardous waste management is embodied in the key piece of federal legislation that governs hazardous waste in America, RCRA. Passed in 1976 and significantly amended in 1984, the Resource Conservation and Recovery Act virtually created the hazardous waste management field. Understanding the provisions of RCRA is an essential first step for anyone interested in entering the hazardous waste management field.

HAZWASTE VOCATIONAL TRAINING

A thoughtful review of the six hazardous waste management steps above produces an interesting observation. A huge number of the nation's ongoing hazardous waste management jobs involve record keeping, requesting and/or granting government permits, routine monitoring, transportation, labeling, maintaining databases, filling out forms, and sharing information. Much of this work can be performed by people with little scientific background but with strong training in the process of management and compliance.

Joe Johnson, Harza Environmental
Earth Worker

Joe Johnson is senior environmental engineer with Harza Environmental Services, Inc. Based in Chicago, Harza is an engineering consulting firm specializing in providing services to municipal, industrial, and government clients in the areas of sanitation engineering and hazardous waste management. Most of Joe's projects deal with water systems and wastewater systems management.

Back in 1978 Joe was part of an SCA high school work group that was doing trail construction work at Great Smoky Mountains National Park in Tennessee. "My SCA experience confirmed that I wanted to get involved in environmental and resource management issues," says Joe. He also adds that he "enjoyed the hard work and established friendships that continue today."

Joe received his B.S. in civil engineering from the University of Notre Dame and his M.S. in environmental engineering from the University of California at Davis, and then worked at an engineering consulting firm before going to Harza seven years ago.

continued

This fact has not gone unnoticed. Throughout the community of hazardous waste generators, the agencies that regulate them and the private firms that support them, the field has begun to generate recognizable job titles with relatively standardized qualifications and knowledge requirements. This, in turn, has fueled the growth of "hazardous waste coordinator" and "hazardous waste technician" certification and/or training programs, which play an increasingly important role in the preparation of hazardous waste professionals. Two-year, vocational, and community colleges are leading the way in this growing area.

This important trend aside, the hazardous waste management field is primarily a scientific and engineering endeavor.

continued from page 121

One reason that Joe opted to work within the private sector is that he finds more opportunity for person-to-person contact and involvement in the planning and design of facilities.

Joe's position at Harza has afforded him the opportunity to study the operation of water systems and wastewater systems in other parts of the world. At present, he is involved in a project designed to assist the government of Panama.

AMERICA'S TOXIC LEGACY: SUPERFUND AND OTHER CLEANUP EFFORTS

When most Americans think about hazardous waste, they think about technicians in moon suits "cleaning up" a Superfund site. Superfund (formally known as the Comprehensive Environmental Response, Compensation, and Liability Act, or CERCLA) was passed in 1980 with the expectation that there were a few hundred problem sites in the nation and that the $1.6 billion appropriated would go a long way toward solving the problem.

By 1990, the EPA had identified some thirty-three thousand hazardous waste sites! Twelve hundred of these made their way onto the government's "National Priority List," and eleven thousand awaited further investigation. It was clear that the taxpayers' initial $1.6 billion was just a tiny down payment on the final cleanup price. Actual cleanups were (and still are) very, very few.

Although the pace of cleanup has accelerated in the 1990s, it is clear that the United States will be identifying and remediating hazardous waste sites throughout the course of our professional lives.

This would be so even if it were not for the massive cleanup activities of the Department of Defense (old military bases and facilities) and the Department of Energy (atomic weapons production sites and nuclear energy plants). Add to this the thousands of contaminated sites that were not bad enough to deserve further federal action and the hundreds of potentially hazardous landfills that will fill up and close down over the next few years; it is evident that hazardous waste cleanup is a secure career choice well into the next century.

The hazardous waste cleanup field requires a workforce that is overwhelmingly dominated by technicians, scientists, and engineers. Grant Ferrier, editor of the *Environmental Business Journal*, has written that hazardous waste management is "perhaps the most technically demanding field to get into."

A look at the Sunday newspaper's classified ads anywhere in the nation will show that he is right. The industry has a particularly strong demand for chemical engineers, environmental engineers, geological engineers, earth scientists, chemists, hydrogeologists, industrial hygienists, computer modeling specialists, toxicologists, lawyers, biologists, and "risk assessment" specialists.

In 1994, the hazardous waste cleanup field is in an exciting state of ferment, growth, and change. The race is on to develop new technologies to remediate contaminated soils and groundwater. Researchers and entrepreneurs are proposing mechanisms such as toxics-gobbling microbes or superheating methods that seal hazardous waste into glasslike substances that can be safely disposed of. New ideas for sealing and lining old dumps or incinerating wastes in ways that do not generate toxic ashes and emissions are being studied. Even basic engineering practices for the removal of toxic soils are undergoing constant change.

The field is still searching for safe, convenient, cost-effective, and permanent hazardous waste cleanup methods. There is a lot of room at the employment table for imaginative and talented scientists and engineers.

THE NEXT WAVE: TOXIC USE REDUCTION AND POLLUTION PREVENTION

Until recently, the hazardous waste management world focused on "controlling" toxic wastes, "monitoring" them, and cleaning up the mess. The huge costs and obvious inefficiencies of this approach have led to a long overdue attack of common sense. Today, the word is out: Reducing the use of toxics and preventing pollution is the preferred way of dealing with hazardous waste.

Although "end of the pipe" and cleanup solutions will be with us for a long time, the new approach is already creating a small revolution in the hazardous waste employment picture. The new heroes are process engineers, chemical engineers, materials experts, chemists, and others who are studying ways to replace hazardous materials with nonhazardous ones or ways to redesign basic industrial processes to prevent pollution. People who can make meaningful contributions to this activity will have no shortage of job offers in the future.

ADVICE FROM THE PROS

If the hazardous waste management field sounds interesting to you, here is what to do next. Identify and talk to people who are solving problems creatively. These people are not hard to find. Just start asking. If a little background would help, start with a simple computer search at a good research library under headings such as "hazardous waste management," "Superfund," "toxic use reduction," and so forth. Read newspaper, magazine, and trade journal articles, and pick out names of people who are quoted. Speak with people from the full range of employers—consultants, activist groups, government agencies, and manufacturers.

Really listen to what these problem solvers say. Where do they see the field headed? What do they advise you to study? What combination of science, engineering, management, and communication is most needed? What do they read to stay on top of the field?

Following these two simple steps will give you a wealth of practical, timely, accurate information from people who are at the cutting edge of the field. After just a few discussions, you will be well on your way to a great career.

For more information on hazardous waste management careers, write to the Hazardous Materials Control Research Institute, 7237-A Hanover Parkway, Greenbelt, MD 20770-3602, (301) 982-9500, or the Air and Waste Management Association, P.O. Box 2861, Pittsburgh, PA 15230, (412) 232-3444, or contact the U.S. Environmental Protection Agency regional office nearest you and the hazardous waste control office within your state government's environmental protection agencies.

For a detailed overview of hazardous waste jobs and a listing of some good colleges, vocational programs, and certification training outfits, order a copy of *Opportunities in Waste Management Careers,* published by VGM Career Horizons. To obtain this book, contact NTC Publishing Group, 4255 West Tonhy Avenue, Lincolnwood, IL 60646-1975.

For paid internships in the hazardous waste management field, apply to the Environmental Careers Organization, 286 Congress Street, 3d floor, Boston, MA 02210.

Chapter Eight

"GREEN-HOT" JOBS

A s the environmental field broadens, new kinds of jobs are created every day. *Earth Work* magazine reports on jobs that are up-and-coming not only to give our readers an edge over the competition, but also because learning about others' work in the environmental field helps everyone. A 1992 *Earth Work* list of "green-hot" jobs made newspaper headlines across the country. Since that time, those jobs have become better known and others have emerged. Some jobs are hot because they are new, while others are hot simply because the supply of workers does not meet the demand.

Perhaps a third of those currently working in environmental and natural resources careers are in well-publicized jobs in solid waste and resource recovery or recycling, so *Earth Work* focuses here on lesser-known opportunities. Because many job seekers are in search of natural resources management jobs, the "green-hot" list includes jobs and careers with growth opportunities. In addition, the list includes some ideas for liberal arts as well as science majors and for entrepreneurs as well as those who prefer to work for an established nonprofit or government agency.

Earth Work magazine's original 1992 "green-hot" list follows:

1. *Environmental manager*

2. *Environmental engineer*

3. *Ecopreneur (e.g., recycling)/ecotourism specialist*

4. *Green marketing manager*

5. *Ecological scientist (esp. wetlands)/biologist/earth scientist*

6. *Fund-raiser*

7. *Environmental lawyer*

8. *Environmental educator and communicator*

9. *Recreation, wildlife, and natural resources manager*

10. *Geographic information systems specialist*

In November 1992, *CBS News* added an eleventh "green-hot" job when it aired its list of "hot jobs" in all fields. CBS put environmental accounting at the top of its list because "companies need to know the costs and benefits of environmental issues." And, when *U.S. News and World Report* listed its "Hot Tracks in 20 Professions" on November 1, 1993, three of *Earth Work*'s green-hot jobs made the list:

- The hottest track in engineering is environmental. Estimates put the number of environmental engineers at only 25,000 to 50,000. *U. S. News and World Report* says the current number is far less than will be needed for such projects as cleanup of contamination of soil and water and design of pollution control systems.

- The hottest track in law is also environmental. In fiscal year 1992, the Justice Department indicted 191 companies, compared with 40 in 1983; some 1,200 contamination sites remained on the EPA's priority list in 1993. Environmental lawyers work for government, nonprofits, private firms, corporate clients, and insurance companies.

- In tourism, travel agents are the hottest track, and one of the hottest niches for them is ecotourism: "Those in other niches, such as ecotours or bike tours, will benefit from the increasing popularity of adventure travel," the article reported.

EARTH WORK'S 1994 "GREEN-HOT" JOBS

1. ENVIRONMENTAL MANAGER. Environmental and occupational safety legislation as well as public pressure for cleaning up the environment have created a much greater demand for environmental managers than there is supply. These managers work for industries that must control environmental practices as well as for recycling industries. The manager identifies risks to the public and to workers from emissions and implements pollution control and safety regulations. Membership in the American Industrial Hygiene Association has blossomed. Training begins with an engineering or science degree; some schools now offer environmental management degrees. Most environmental managers have advanced degrees and can earn $50,000 starting out, with more than $100,000 going to upper-level managers at Fortune 500 firms.

2. ENVIRONMENTAL ENGINEER. Environmental engineers are needed at engineering firms, in state government, at the U.S. Environmental Protection Agency, and at companies around the nation. Starting salary with a B.S. is around $31,000; with an M.S., $40,000. Both chemical and civil engineering schools produce "environmental engineers" with cross-training in the biological sciences. Air quality engineers, who are in charge of analyzing and controlling air pollution under new Clean Air Act regulations passed in 1990, often have mechanical engineering degrees if involved with power production. A chemical engineering degree is attractive in many environmental professions; some twenty-two thousand engineers will be needed to clean up toxic chemicals alone. In the bioremediation field, which uses bacteria to destroy chemicals, an entry-level bachelor's degree in chemical environmental engineering can get $38,000; combined with a master's in civil-sanitary engineering, $42,000. See the books on environmental engineering by Nicholas Basta in the Recommended Book List.

3. "ECOPRENEURS" AND/OR ECOTOURISM SPECIALISTS. Environment-related products and services present many business opportunities for entrepreneurs in recycling, conservation, safe foods, and entertainment. For example, as the United States continues to move toward a service economy and a growing number of communities see the economic value of expanding tourism and protecting their environment, ecotourism will expand. State agencies and private companies will increasingly seek out people with academic training in both park and recreation management and tourism science. Ecotour companies employ biologists and naturalists as tour guides; some of these professionals have established their own companies or become travel consultants. Meanwhile, nonprofit organizations support scientists performing research that will attract tour clients who want to contribute to environmental projects at attractive sites. *Ecopreneuring* (see the Recommended Book List) is a start-up guide for entrepreneurs.

4. GREEN MARKETING MANAGER. M.B.A.s with environmental specialization can call the shots on their salaries because America's large corporations can make or break their businesses by how they respond to environmental problems and by whether the public views their products as "green." At New York University's graduate business school, environmental studies have been incorporated into marketing and product development courses. NYU career counselor Sue Cohn advises those with bachelor's degrees in marketing to apply as assistant marketing managers at starting salaries of $20,000–30,000. M.B.A.s can get $45,000 to $55,000.

5. ECOLOGICAL SCIENTIST/BIOLOGIST/EARTH SCIENTIST. Biological studies open the door to many environmental professions, including ecology and fish and wildlife management and research. *Earth Work* magazine lists numerous job openings in this area. Although many ecologists are involved in research, new applied opportunities will open up for ecological scientists and/or biochemists in evaluating damages in natural resource damage suits brought by the Department of Interior or other "natural resource trustees" under Superfund. One of those trustees is the National Oceanic and Atmospheric Administration

(NOAA), which also hires earth scientists, including those with expertise in geology, oceanography and marine science, soil science, and atmospheric science. The latter is experiencing a boom because of the concern over global warming. Wetlands ecologists are in demand because of the need to curb the rapid disappearance of wetlands nationwide, especially in coastal areas that support fisheries.

The Ecological Society of America reported in the mid-1980s that starting salaries were about $20,000 for biologists with a bachelor's degree; Ph.D ecologists could begin at more than $50,000. Ph.D geoscientists may start at $36,000 in government and $50,000 in industry.

6. FUND-RAISER. With the recession cutting back on nonprofit memberships and foundation support, directors of development in nonprofit environmental organizations command almost as much as CEOs—an average of $57,000, with some development directors making $125,000. This is an area of opportunity for liberal arts or business majors, but do not expect to start as a director. A fund-raising internship or entry-level salary in the $18,000–$25,000 range might double within five years for those who demonstrate fund-raising, public relations, computer, and marketing savvy as well as environmental commitment on the job.

7. ENVIRONMENTAL LAWYER. Environmental law is the fastest growing legal specialty. Environmental lawyers work at local, state, and federal agencies, in corporations, and as consultants with advocacy groups. The need has increased as the volumes of environmental regulations expand. Starting salaries may range from $30,000 at a nonprofit to $80,000 at a corporation.

8. ENVIRONMENTAL EDUCATORS AND COMMUNICATORS. The National Environmental Education Act of 1990 has opened up new opportunities and EPA grant monies for environmental education. Environmental educators are needed in K–12 classrooms across the country as well as in curriculum development at educational institutions and consulting firms. They also work at colleges, nature centers, parks, zoos, and museums. Environmental communications is becoming a profession unto itself, with graduate school programs available in environmental journalism and even environmental filmmaking. As the environment becomes big news, companies are hiring community relations managers, and cities are hiring recycling coordinators who serve in a public relations role. Salaries vary widely according to place of employment.

9. RECREATION, WILDLIFE, AND NATURAL RESOURCES MANAGERS. Besides the National Park Service, U.S. Forest Service, and U.S. Fish and Wildlife Service, lesser-known agencies such as the National Marine Fisheries Service and the Bureau of Land Management offer fantastic working experiences and new jobs. In agencies such as BLM large numbers of older managers are soon due for retirement, creating new openings, and the NMFS is expanding. Academic backgrounds of employees include majors in wildlife, parks and recreation, fisheries, or natural resources management. Starting salaries are around $23,000 but go up to high-level government grades.

10. GEOGRAPHICAL INFORMATION SYSTEMS SPECIALIST. Employees with graduate degrees in forestry, parks, or natural resources fields who specialize in geographical information systems (GIS) are in great demand at public and private agencies. GIS computerized systems increasingly are used by national and state parks and forests, the Corps of Engineers, Indian reservations, big private companies, family farms, tourist developments, and others to capture and analyze a variety of natural resources data.

The February 1994 issue of *Geo Info Systems* magazine explains more about how GIS technology works and calls it indispensable in protecting wildlife in the Space Age. At Kennedy Space Center in Florida, NASA scientists use geographical information systems to monitor the effects of launches on wildlife at the Merritt Island National Wildlife Refuge. For example, physical geographers, ecologists, and GIS specialists compile databases on endangered species at the refuge and "overlay" them with databases on the effects of launches. Another example is overlaying sightings of wildlife such as manatees with LANDSAT satellite photos. The technology also enables them to monitor episodic environmental impacts such as those of shuttle exhausts.

CAREERS IN GREEN MARKETING

A career in "green" marketing should be approached the same way you would any traditional marketing career—by obtaining the functional marketing skills and relevant work experience. It is your marketing skills, not your environmental background, that will determine your hiring potential in the marketing field. Possessing an environmental background in addition to your marketing skills can only add to your credentials in pursuing a position in green marketing.

Choices in green marketing activities run the gamut from basic services and products to marketing approaches. Green marketing involves asking questions about what makes a product green and what environmental issues can be addressed in the marketing and designing of a product, from the extraction of raw materials to packaging, distribution, and reuse. For example, when marketing by direct mail, one green consideration might be the amount of waste involved in direct mail as compared with print media. If using catalogs, you might evaluate the best ways to lessen the waste: using recycled paper, making the catalog lighter in weight, or using an entirely different medium such as radio or television. As a green marketing manager, you are responsible for combining the traditional marketing techniques with an environmental screening process.

Michael Darling, clinical professor of marketing at New York University's Stern School of Business, creator of Earthword, an environmental education game, and consultant to companies in many different industries, issues a cautionary ethical note on the screening process: "Marketing with an eye to the environment can be used to create a competitive advantage; however, too often this competitive advantage is sought through claims rather than fundamental product redesign. This can be expected in the short term and is possibly good marketing. But sooner or later, the more basic issues will need addressing.

"Several leading corporations," emphasizes Darling, "are taking this broader view and, while seemingly unresponsive to the issues today, are busy innovating for the future. We will see a plethora of greener goods and services over the next five to ten years, and the mainstreaming of the environment will have to come to fruition."

He explains that people with a leaning toward green marketing should become competent marketers first and then build their environmental platforms, particularly because greening can be expected to become so integrated into professional business practices. "If green is your goal, establish your knowledge base through reading, and build your network of environmental contacts; . . . you'll soon be in a powerful position to influence more rapid movement toward a more environmentally conscious business world," Darling says.

To gain an edge in establishing your knowledge base, the following publications will be most useful in greening your marketing career.

> Green Is Gold *(1992) by Patrick Carson and Julia Moulden (Toronto: Harper Business), $19.95.*
>
> Keeping Your Company Green *(1990) by Stefan Bechtel (Emmaus, PA: Rodale Press), (215) 967-5171, free.*
>
> The Green Consumer Letter *and* The Green Business Letter *(Tilden Press, 1526 Connecticut Avenue NW, Washington, DC 20036), (800) 955-GREEN. You can also request copies of Makower's book* The Green Consumer *and other book titles with an environmental focus.*
>
> The Green Consumer *(1993) by Elkington, Hailes, and Makower (Island Press, Box 7, Dept. 4C2, Covelo, CA 95428), (800) 828-1302. Write or call for their catalog of environmental books.*

BIODIVERSITY CONSERVATION CAREERS

Since the 1992 Earth Summit in Rio de Janeiro, biodiversity has come into vogue. But what exactly does this popular catchword mean? Under the biodiversity umbrella are genetic diversity, species diversity, and ecosystems diversity. Genetic diversity is the variation of genes within a species. As described in a World Resources Institute report, genetic diversity allows plants and animals to evolve and adapt to new conditions. Species diversity refers to the number and variety of species within a region. Ecosystems diversity is harder to measure, but the term describes the associations of species within a geopolitical or natural region. For example, the interaction among species in a region perpetuates the carbon and nitrogen cycles and establishes the food chain. If one species (or sometimes even one subspecies) is lost, all dependent species are threatened.

We cannot fully comprehend the intricate webs of interdependence that link species, because so many species remain to be discovered and their roles elucidated. Each day, an estimated 140 species of plants and animals become extinct in tropical rain forests. And with their loss, we forfeit the benefits that those 140 might have afforded humankind.

Some argue that biodiversity must be conserved even at the expense of economic gain. In practice, however, the reverse seems to be true. Rain forests are slashed and burned to clear land for farming. Old-growth forests are cut for timber. Habitats of endangered species are destroyed to make room for development. Insect life is decimated by the application of agricultural pesticides. Unfortunately, only in recent years have we awakened to the incalculable loss of biological diversity caused by human activity.

Both conservationists and developers have traditionally viewed biodiversity and economic growth as a trade-off: One can be sustained only at the expense of the other. But a new, revolutionary approach that is emerging maintains that by using certain economic incentives, one may preserve biodiversity and create jobs at the same time. In fact, proponents of this approach say economic development depends upon conservation, and the two should be advanced together.

A well-known advocate of this common-ground approach is Edward O. Wilson, Harvard University biologist and author of *The Diversity of Life*. According to Wilson, "Except in pockets of ignorance and malice, there is no longer an ideological war between conservationists and developers. Both share the perception that health and prosperity decline in a deteriorating environment." One reason is that "useful products cannot be harvested from extinct species. If dwindling wildlands are mined for genetic material rather than destroyed for a few more board feet of lumber and acreage of farmland, their economic yield will be vastly greater over time."

One example of the common-ground approach is plant-derived medicines. Some cultures have exploited the pharmacologic activity of certain plants for centuries. In South America, the bark of the cinchona tree was found to cure malaria as early as the fifteenth century. Some years later quinine was extracted from the bark and identified as the effective substance. In 1958 investigators discovered that the rosy periwinkle, *Catharanthus roseus* of Madagascar, produced vinblastine and vincristine, effective treatments for Hodgkin's disease and acute lymphocytic leukemia. Taxol, which the National Cancer Institute currently claims is "the most promising cancer treatment in twenty-five years," was first isolated from the Pacific yew tree of Oregon and Washington. Even though the honor roll of naturally derived compounds is already long, researchers are only beginning to understand the vast medicinal potential of naturally occurring compounds.

The National Cancer Institute, pharmaceutical companies, botanical gardens, and universities are pioneering programs to identify natural products and bring them to the marketplace. As part of these efforts, the National Cancer Institute, in collaboration with the Missouri Botanical Gardens, the New York Botanical Garden's Institute for Economic Botany, and the University of Illinois, is conducting natural products research in Africa, tropical Asia, and Latin America, respectively.

Dr. Jim Miller of the Missouri Botanical Gardens in St. Louis sums up his project: "Our institution is involved only in collection and screening of plant materials. We collect the basic botanical data, and we provide the data to conservation organizations to use when developing conservation programs." Miller's Natural Products Research group collects and identifies botanical samples. The samples are sent to a chemistry laboratory where they are tested for pharmacological activity. Miller says that once the project chemists identify a promising substance, known as the lead compound, his group's involvement ends. At that point, the pharmaceutical companies take over and embark on "high tech" pharmacological research. In most cases, according to Miller, naturally occurring compounds have undesirable side effects. Chemical modification of the compound is usually necessary before human clinical trials can begin.

"There is a real shortage of well-trained field biologists," says Miller. "We need people with botany degrees and field experience, who are familiar with the language, customs, and way of life in the regions where we are working [Africa]." The best way to gain field experience is through doctoral research; Miller spent a total of five months performing field research in Central and South America while working toward his doctorate. "Fluency in French or Spanish is necessary for our African studies. Other programs require Portuguese, Chinese, or Japanese," Miller continues. "We are looking for people to go to Ghana, Ivory Coast, and Cameroon."

Another way to gain field experience is through the Peace Corps or other volunteer programs. Miller elaborates, "Of the fifty or so Ph.D.-level biologists in our Natural Products Research group, five or six came from the Peace Corps."

The merging of biodiversity preservation and economic benefit is illustrated by an innovative agreement reached last year between the National Institute of Biodiversity of Costa Rica (INBio) and Merck and Company, the world's largest pharmaceutical company. Under the agreement INBio provides Merck with plant and insect samples, an arrangement that, in the words of a Merck spokesperson, "gives Merck access to the amazing collection of biological specimens found in Costa Rica, while working with an organization that will protect and conserve the habitats from which these samples are drawn." Costa Rican scientists will process the samples, and Merck laboratory personnel will evaluate them for pharmaceutical or agricultural applications.

Technology transfer is an important aspect of the Merck-INBio agreement. Merck donated sample-processing equipment to the University of Costa Rica and trained Costa Rican scientists. In addition, an exchange program for Costa Rican and Merck scientists was established.

Most important, the agreement provides economic incentives that will help sustain the rich biodiversity of Costa Rica. Merck will pay INBio $1 million to cover the cost of conservation activities, and INBio will receive royalties on sales of any products developed from these samples. A report by the World Resources Institute estimates that "even if INBio receives only 2 percent of royalties on pharmaceuticals developed from Costa Rica's biodiversity, it would take only twenty drugs for INBio to be able to earn more funds than Costa Rica currently gets from coffee and bananas—two major exports."

The Merck-INBio and Missouri Botanical Gardens projects promote nondestructive uses of the rain forest and show that economic benefits and conservation can share common ground. Projects like these demonstrate that a balance between human progress and long-term sustainability of the world's biotic wealth can be achieved. As natural products research expands, more people will need to be trained in biodiversity conservation, and increased financial and intellectual resources will need to be devoted to critical fieldwork.

INTERNATIONAL ENVIRONMENTAL CAREERS

Wildlife populations, air currents, and rivers have always crossed national boundaries; now environmental professionals are doing so in a big way. The 1992 Earth Summit was a turning point for international environmental careers. Converging on Rio de Janeiro, Brazil, were 118 heads of state, thousands of environmental and conservation professionals from citizens groups and government agencies, and tens of thousands of other citizens from the 180 nations represented. It was the largest meeting of heads of state in history and attracted nine thousand reporters, the largest media group on record. The summit events included the official United Nations Conference on Environment and Development (UNCED) as well as the Global Forum—a massive "people's assembly." More important, the Earth Summit raised the consciousness of the world about environmental challenges and resulted in Agenda 21—a nonbinding but comprehensive blueprint to protect the environment and encourage sustainable development. It also produced treaties to limit emissions of greenhouse gasses and to protect biodiversity. Although President Bush refused to sign the latter treaty at the summit, President Clinton signed the biodiversity treaty on Earth Day 1993.

The Earth Summit was the culmination of two years of international negotiations; it also marked the start of a process, not the end result, because the biodiversity treaty, for example, still had to be signed by the president and ratified by the U.S. Senate. Thus, many conservation professionals are still working on the issues that came before the Earth Summit, as well as on other international issues.

A number of international treaties or conventions already attract oversight and participation by these professionals, including the Convention on International Trade in Endangered Species (CITES). In addition, the debate over dropping trade barriers through

NAFTA and GATT during 1993 and 1994 drew the participation of most of the major environmental organizations. At stake were U.S. environmental laws and the import restrictions they set—such as refusing imports of tuna from Mexico that have been captured by unsafe methods in nets set around dolphins.

The United Nations first addressed environmental issues on a large scale at the 1972 U.N. Conference in Stockholm. It resulted in the creation of the United Nations Environment Programme, the body that now organizes worldwide environmental monitoring, spotlights emerging environmental issues, and catalyzes negotiations that yield new international environmental law. Twenty years later, the much larger 1992 Earth Summit signified an unprecedented integrated approach by the United Nations by publicly recognizing that environmental issues are interrelated and inseparable from development strategies. This recognition will inevitably change the structure of current international institutions, reverberating in policy making and investment, and will broaden the range of future careers, in which both technical environmental knowledge and international relations skills will be necessary.

Those currently working in international environmental careers can be found at nonprofits, in the government, and in private industry. They might work for the United Nations Environment Programme, in international nonprofits such as the World Wildlife Fund and World Resources Institute, or in national environmental organizations with international programs such as the Natural Resources Defense Council, National Audubon Society, and National Wildlife Federation.

Government agencies such as the State Department, Environmental Protection Agency, National Park Service, U.S. Fish and Wildlife Service, and National Oceanic and Atmospheric Administration have international offices to deal with treaty obligations and global issues in their fields. In the private sector, international development projects are requiring more environmental expertise from engineering and consulting firms to become sustainable.

Many conservationists have moved into international positions from other nonprofit environmental or government work, but new jobs are likely to open up in years ahead. Larry Williams, director of the Sierra Club's International Program, explains, "It's an evolutionary process. Groups national in scale will gravitate to international as they see global connections. . . . It's inevitable that careers in the international environmental field will expand." The Sierra Club's international involvement began twenty years ago and has grown under the foresight of Sierra Club chairman and former executive director Michael McCloskey.

Interviews with professionals in the international environmental field outline a number of general trends that will demand more expertise, and more jobs, in the international environmental sphere.

TREND 1: BETTER INTERNATIONAL POLICY IS INCREASING THE NUMBER OF INTERNATIONAL JOBS. Those involved directly in the UNCED process see breakthroughs in United Nations policy as the first step toward progress in the international environmental arena. Hilary French, senior researcher for Worldwatch, asserted before the Earth Summit, "How strong an impact any of the UNCED conventions will have on employment and other trends will depend on the treaties themselves. There are large employment opportunities, especially in energy efficiency and renewable energy. The pace of employment depends on how quickly we move to negotiate this process into motion."

Gareth Porter of the Environmental and Energy Study Institute says that "there will be a need for specialists in sustainable development—in sustainable economics, forestry, and agriculture, for example. More organizations must become involved in international sustainable development issues than 'pure' international environmental issues. U.N. institutions—United Nations Development Program and United Nations Environment Programme—will most likely expand."

TREND 2: ENVIRONMENTAL JOBS ARE MOVING FROM A SINGLE-ISSUE FOCUS TO AN ECO-SYSTEM APPROACH. UNCED reflected the need to coordinate multifaceted issues of environment and development. As the human ecosystem is becoming recognized as global, national organizations are increasingly finding themselves working on international issues. Fran Spivy-Weber, director of Audubon's International Program and chair of administration of U.S. Citizens Network on UNCED, explained that, "Audubon is involved in UNCED in response to the growing international interests of its members. Ironically, the biggest areas of employment for environmentalists, especially international ones, are New York City and Washington, D.C. You can work on international environmental issues no matter what your background, and this is reflected by the variety of groups at UNCED."

On this same subject, Susan Weber, executive director of Zero Population Growth and treasurer of the Global Tomorrow Coalition, says that "with UNCED, we are finding ourselves needing more and more to communicate around the world. The future, sooner or later, is in environmental and population endeavors. Whether our government wants it or not, that is where the enterprise will be." And Matthew Gianni, a fisheries campaigner for Greenpeace International, says that whether hired by the international office or one of the national ones, "Eventually everyone works on an international project." A former commercial fisherman who works on marine pollution projects, Gianni notes that "coastal fisheries worldwide need liaisons, to ensure their development doesn't occur at the expense of ecosystems. One needs broad-based expertise, not only fisheries-specific knowledge. The field needs more people with the ability to react from an international perspective in negotiations and networking."

TREND 3: INTERNATIONAL DEVELOPMENT PROJECTS WILL REQUIRE MORE STAFF IN MONITORING, TECHNOLOGY TRANSFER, SOCIAL SCIENCES, AND ENVIRONMENTAL ASSESSMENT. Barbara Bramble, director of international programs at National Wildlife Federation, predicts "a future for small research and development firms that team up in joint ventures with corporations on projects such as power plant efficiency." She maintains that sustainable development should be on the same scale as traditional development projects: "If it were seen that way, there would be plenty of jobs available in international environmental fields, and it is inevitably going that way. Within multilateral banks, there will be a lot of internal cleanup to do on past development projects." She says NGOs could be created to monitor for the banks, to make banks publicly accountable, "but only if they will commit to bank monitoring full-time." Bramble also predicts more environmental assessment jobs within banks. To assess the needs of indigenous peoples and make development projects sustainable, "banks will need to hire social scientists familiar with language and culture to work in the field."

David Hulse, Asia program officer for the World Wildlife Fund, says, "Although the worsening economy has deflected some attention away from environmental issues . . . career opportunities for environmental professionals remain positive. What I find particularly important is the new emphasis that bilateral funding agencies such as U.S. Agency for International Development and multilaterals such as UNDP and the World Bank are placing on environmental programs. Not only are these agencies taking steps to mitigate the environmental impacts of their traditional development projects, they are also soliciting the input of conservation professionals in program design, particularly through the Global Environmental Facility."

TREND 4: THE ENERGY LINK TO SUSTAINABLE DEVELOPMENT IS BECOMING MORE VISIBLE. "Global attention to the need for energy efficiency in developing countries as a pollution prevention strategy promises to be greater than ever before," according to Deborah L. Bleviss, president of the nonprofit International Institute for Energy Conservation (IIEC). She says the energy issue "has just come home to roost" because of publicity about global warming; the capital crisis in developing countries, which has spurred interest in energy efficiency and in getting more energy services for less money; and local opposition to energy projects because of pollution and disruption.

Founded in 1984 on the premise of fostering sustainable development through energy efficiency, IIEC has grown: Its budget has blossomed, and its staff has expanded from one to twenty in that time. The IIEC acts as a "broker" or facilitator between institutions with experience in energy efficiency technologies and policies—generally located in industrialized countries—and those in developing countries with the need for such experience. From its headquarters in Washington, D.C., and offices in Thailand, Chile, and soon in Belgium, IIEC stimulates information dissemination, demonstration projects, and development of local energy conservation business activity.

Bleviss does not predict an immediate surge of new international environmental jobs but a steady growth over the next decade. She points out that the educational and scientific communities need to adapt to the demand.

TREND 5: INTERNATIONAL ISSUES ARE REQUIRING NATIONALLY AND CULTURALLY DIVERSE STAFFS AND LOCAL EMPOWERMENT. Many conservationists in the international arena stress that conservation techniques learned in the Northern Hemisphere do not work when imposed on the Southern Hemisphere and that "northern" conservation professionals must focus their efforts on the exchange of technical expertise rather than a one-way sharing of ideas. They also note the need for cultural diversity and for language training. Greenpeace's Gianni says, "Potential employees from northern countries ideally should be multilingual, with insight on the perspective of southern countries."

Michael Philips, program manager for policy and development institutions at IIEC, points out the need for political, as well as cultural, sensitivity: "The ideal person for the international environmental job would be a technology-minded person who can combine those skills with people management skills and the ability to implement programs. We have the technology for large-scale sustainable development right now, but the field needs people who are politically adept" and can understand the many institutional and political obstacles that become barriers to implementing the technology today.

TREND 6: NEW EDUCATION AND SKILLS ARE BEING EMPHASIZED AS ENVIRONMENTAL TRENDS GO GLOBAL. Joan Martin-Brown, the United Nations Environment Programme's representative to the United States and the diplomatic community, notes, "Education is particularly important for the environmental community because the field requires a lot of matrix thinking"—for example, combinations of environment, science, law, and education. The interdisciplinary nature of the field means that there is no single preparation for an international environmental career, but, she says, "It's imperative that you have a second language in depth. American young people are competing in a global market with people who have at least two or three languages interchangeably. In addition, you need technical or legal training."

The broad spectrum of disciplines necessary in the international environmental area includes law, management, and public policy making and implementation. Communication skills, especially writing skills and foreign languages, are in high demand already.

According to IIEC president Deborah Bleviss, "Graduate schools have not kept pace with the demand for professionals with training in more than one discipline. For example, there is a need for economists with energy training. Successful implementation of energy efficiency and renewable energy projects requires the knowledge not only of physicists and technicians but also of behavioral and social psychologists, because energy patterns involve changing people's behavior. For the next five to ten years, Bleviss advises those interested in international careers to devise their own programs.

Finally, it is important to note that the best jobs one can have in the international environmental field fifteen to twenty years from now may not even exist today. But, in order to solve environmental problems that are crossing national boundaries and affecting international economies, today and tomorrow, people with combined environmental knowledge and international relations skills are becoming indispensable. Those who gain the additional skills to meet the international job challenges in the environmental arena, including cross-cultural skills, political negotiating, communications, and language skills—will be that much further ahead when the new jobs are ready to be filled or created.

THE GREEN TEACHERS: "THE ENVIRONMENT IS WHERE YOU ARE"

The desire to teach leads to a staggering array of career options. Environmental educators can be nature center staff or park rangers, classroom teachers or advocates for a cause. They may work with city dwellers or country dwellers, preschoolers or senior citizens. "As a group, environmental educators tend to be very optimistic. They believe that education and communication lie at the heart of solving our environmental problems," says Lee DeAngelis, regional director of the Great Lakes Office of the Environmental Careers Organization.

There is good reason to look to the 1990s with optimism. Educators see an American public anxious to clean up the environment. New career opportunities are opening up. A survey conducted by the Environmental Careers Organization estimates that jobs in environmental education and communication will grow by 5 percent to 10 percent per year throughout the decade. Much of this growth will come as those in the field find new ways to target their audiences and messages.

What is more, a new wave of federal support for environmental education is rising. Congress signaled its commitment to the effort with the passage of the National Environmental Education Act of 1990. The act gave the Environmental Protection Agency responsibility for supporting environmental education from grade schools through universities and for encouraging students to pursue careers in the field.

How much larger a role the federal government will play is anyone's guess. The EPA's Office of Environmental Education has launched a number of programs to build partnerships and to set standards. The largest single project is the National Consortium for Environmental Education and Training, a partnership of universities, nonprofits, and corporations developing and supporting programs to help elementary and secondary school teachers to present issues more effectively.

In 1992, the office also administered $2.5 million in grants. But with a full-time staff of only nine employees, much of the office's success will depend on whether it gets additional support from the Clinton administration, says its director, Brad Smith.

Janet Ady, a Fish and Wildlife Service environmental education specialist, observes, "Only recently have we realized that in order to effectively manage wildlife we have to effectively manage people too."

TARGET AUDIENCES

Evidence that environmental education is swiftly moving into the mainstream comes from many sources. Newspapers and television stations have hired environmental reporters. New magazines with a focus on the environment have appeared. At least thirty states are considering some form of environmental education mandate. Also, membership in the North American Association for Environmental Education (NAAEE) and the Alliance for Environmental Education (AEE) is growing dramatically.

Across the board, the emphasis is on localizing environmental issues and making them personally relevant, DeAngelis says. When the National Wildlife Federation, for example, dovetails children's materials with lessons from its *Ranger Rick* magazine, teachers can use the ideas to carry out projects in their school neighborhoods.

Established environmental education programs in major cities also rely heavily on outreach. In Louisville, Kentucky, a school summer camp program works hand-in-hand with nine organizations, from the YMCA to scouts. Community nature centers also serve schools.

In Vermont, Richard Thompson Tucker, executive director of the Merck Forest and Farmland Center, runs a program ambitiously aimed at many audiences. The center owns 2,800 acres in the mountains with twenty-six miles of trails open to hiking and skiing. Family activities, from sheep shearing on a working farm in the spring to snowshoe hiking in the winter, help bring in paying customers. Increasingly, the survival of nature centers will depend on how well they can match their resources to the needs of the communities they serve. Tucker is now looking at designing programs for both local day-care centers and Elderhostels.

According to Dr. David Wicks, coordinator of environmental education for the Jefferson County Public School System that serves Louisville, resources are good. As supplemental training, Louisville teachers may spend a week on a school nature preserve and historic farm then communicate to their students a vision of the past in the context of the environmental present.

TRUMP CARD

In most schools, the model for teaching environmental education calls for integrating lessons into many topics instead of teaching everything as a stand-alone or as part of a unit in biology class. Few elementary and secondary teachers will lead classes solely on the environment, but being able to teach about the environment may be a trump card for new teachers, especially those in the sciences. Environmental education may also get a big boost from movements in science education that stress problem solving and the role of science in society. "It may mean that science teachers coming out of school are going to have to wear their own environmental educator hats," speculates DeAngelis. At the same time, Wicks says, "It's much easier for environmental educators to get a job if their background is tied into a nationally recognized field such as social work or the sciences."

As schools emphasize the environment, educators are calling for information they can trust, thus spurring additional research at the university level. "There is a tremendous interest right now in asking the question, What is environmental literacy? and looking at setting standards," says Bora Simmons, an assistant professor at Northern Illinois University in De Kalb. Enrollments in college courses on the environment are also setting records. "But the problem is that universities are shrinking, so you have offsetting trends," she says. "There is a demand for more college professors, but the question is where the funding is going to come from." In her job, Simmons splits her time between classroom teaching, researching different ways of motivating people toward environmental change, and working outdoors at the Lorado-Taft Field Campus. The campus, a 150-acre site forested with pine and oak along a river, hosts visitors from third graders to graduate students for three- to five-day programs.

The question of how far you can go in motivating people is also a hot one. "If you just tell people the facts, the idea that they will follow through and do the right thing is not supported by research," warns Edward J. McCrea, executive director of the North American Association for Environmental Education (NAAEE). "The big challenge that environmental education faces is to get people to want to act." As the field develops, more emphasis is being placed on students' skills in problem solving and conflict resolution. At the same time, organizations producing materials for schools stress ways in which kids can make up their own minds.

WORKING WITH CORPORATIONS

A path that environmental educators walk more frequently involves work with corporations. Increasingly, corporations hire environmental communicators to work with regulators and to promote their company's views to the public. Others support development of environmental education materials.

For college graduates interested in exploring corporate jobs, DeAngelis suggests looking at where top management in a particular company stands on the environment. "In the most successful companies, sensitivity to environmental concerns is part of their mission. You look to see that they are backing up their words with actions and not just committing guilt money," DeAngelis says.

Many corporations have also entered into cooperative agreements with nonprofit groups to meet specific goals. For example, McDonald's Corporation joined forces with the Environmental Defense Fund to decide to replace the polystyrene packaging used in McDonald's restaurants with wrappers and bags made of recycled paper.

Another example is the Wildlife Habitat Enhancement Council, a joint venture between corporate and conservation communities to invest in environmental protection and habitat enhancement. Programs include creating pockets of habitat in industrialized zones; providing cover and shelter for migratory birds; setting up buffers for wild turkey, fox, and deer; and using right-of-way management techniques for wildlife through reduced mowing techniques.

As a workplace for educators, a corporation is still by far the exception, with 60 percent of educators working in the public sector and 30 percent in nonprofits. Business skills, however, are often prized, especially among nonprofits, where organizations like nature centers must develop creative approaches to fund-raising.

DISCOVER YOUR NICHE

The environmental education field is still one, however, whose career ladder does not come in a handy off-the-shelf model. Flexibility is absolutely essential.

Sonya Wood, for example, a marine extension agent for the Florida Sea Grant Program, is one of those whose ideal career did not land on the doorstep the day after graduation. "When I got out of college, the closest I got to a job as a marine biologist was working as a waitress in a seafood restaurant," she says. Facing an almost impenetrable job market, Wood went back to school, earned her teaching certificate, and taught for a year in the classroom before discovering her niche in the space between research and teaching.

"Most people know what they love, and they should pursue it," says Wood, a self-confessed beach bum at heart. "But finding out how far they can go may take a little time." Today, Wood travels around the state filling the public in on the latest research in marine biology. One day, she might appear in a business suit addressing a county planning board. The next day, she may be wearing jeans and a T-shirt talking to shrimpers down at the dock. Her busy schedule also includes leading plenty of hands-on experiences. She inspires teachers by leading them on a ropes course during a weekend workshop and gives children the chance to babysit nesting sea turtles as part of an adopt-a-shore program.

SEAL YOUR DEAL

Though they walked many paths to their own careers, veterans in the field talk about common stepping-stones or building blocks. Many an environmental educator's interest first began with family hikes or activities like scouting. Volunteer work or internships during college that focus on acting as a leader can pay dividends, come graduation. Among skills at the top of employers' wish lists include a familiarity with business or marketing and the ability to write well.

After you graduate, seasoned environmental educators almost universally recommend two or three years of work experience before going on for an advanced degree, increasingly your ticket to better pay. "Work gives graduates a sense of reality," says Gary San Julian, vice president of research and education for the National Wildlife Federation. "After some field experience, they can go back and get their master's." The experience can also help graduates refine their goals and find mentors. "If people want to build a long-term career, they have to create a vision for themselves," says Frank. "They have to carve out their niche and pursue it with a vengeance."

NEW OPPORTUNITIES IN NATURAL RESOURCE DAMAGE ASSESSMENT

Governmental regulations are targeting hazardous substances, and the bull's-eye is painted on the wallet of every company that has or might have released these substances into the environment. Faced with the prospect of significant monetary damages, government's and industry's demand for professionally prepared Natural Resource Damage Assessments (NRDAs) will probably climb in the near future. This demand will increase job opportunities for a variety of environmental and natural resources professionals.

The Department of the Interior (DOI) proposes a new and far more comprehensive approach for assigning monetary damages when fish, wildlife, forests, groundwater, and other natural resources are injured as a result of a hazardous substance discharge by a potentially responsible party (PRP). By raising the stakes for hazardous substance discharges, DOI hopes to deter some releases. In addition, the awards will be used for environmental restoration. DOI and the National Oceanographic and Atmospheric Administration (NOAA) have been seeking these sorts of damages for some time in the role of "natural resources trustees," and now state governments and Native American communities are becoming more involved in seeking compensation for injuries to the natural resources under their control.

DOI's emphasis on damage awards should trigger an increased demand for NRDAs. An NRDA consists of (1) evaluating the extent of injury or adverse impact to the natural resource and (2) projecting the monetary damages that will be required to compensate for the injury. Although it may seem that the injury to wildlife, habitats, and aquatic life caused by massive releases, like the *Exxon Valdez* oil spill in Alaska, is incalculable, NRDA teams can use existing and proposed regulations to determine damages for injury.

NRDA teams include biologists, oceanographers, meteorologists, physical scientists, chemists, mathematical modelers, natural resource economists, and field technicians. Injury and damage evaluations often involve mathematical models that calculate the extent of natural resources injuries based on data about the release, the ambient conditions, and the affected ecosystems. For example, an NRDA team might use a food chain or physical transport model.

Where natural resources have a known commercial value, injuries can be readily translated into monetary value. This market approach would be used, for example, if a commercial fishery was damaged by an oil spill. But in other cases, damage assessment is more art than science. When it comes to measuring the loss of recreational fishing areas or injuries to wetlands or tidal zones, there is much room for subjective judgment. In these situations, natural resource economists use "contingency valuation theory" to assign dollar values to nonmarketable natural resources. The need for skilled assessments has spawned degree programs in natural resource economics, including those offered by the University of Maryland's Department of Agricultural and Resource Economics and the State University of New York's College of Forestry and Environmental Studies.

Most natural resource trustees and PRPs contract out their NRDA work, and that means new job opportunities within environmental engineering consulting firms. Large firms can pull together an NRDA team from their pool of professionals and get on-site in short order.

DOI's pending regulations will create new opportunities for field scientists and natural resource economists because those regulations threaten to hit industry where it hurts—on their balance sheets.

PROFILE OF AN NRDA TEAM

Robert Dwyer, senior project manager with Environmental Resources Management in Annapolis, Maryland, outlines the types of professionals needed by environmental engineering firms for NRDAs.

PROJECT MANAGER: Assembles the team and directs its efforts in an efficient way. Must be familiar with environmental regulations, statistics and sampling theory, mathematical modeling, and natural resource economics. Usually has an advanced degree in engineering or science backed with extensive experience.

FIELD ECOLOGICAL SCIENTISTS: Count, measure, and apply statistical methods to evaluate injury to involved resource; for example, foresters and wildlife biologists assess injury to a forest. Bachelor's or master's degree in biology, with an emphasis on statistics and sampling theory.

CHEMISTS: Correlate ecological damage with presence of contaminants. Usually work in an off-site laboratory facility. Extract contaminants from soil, sediment, and aqueous matrices and adhere to strict quality control procedures for analysis and documentation. Methods development requires an advanced degree and extensive experience. Analysts usually have a bachelor's or master's degree in chemistry. Lab assistants may be graduates of high schools or junior colleges.

MATHEMATICAL MODELERS: Develop and apply computer models to evaluate effect of contaminant in a given ecosystem. Some models may also estimate monetary damages. Bachelor's or master's level engineer and/or scientist with some experience. Expertise with modeling is often gained through "osmosis," rather than course work.

SCIENTISTS: Depending on the ecosystems affected by the release, hydrologists, geologists, oceanographers, meteorologists, and marine biologists may be needed on the team to measure the impact of the contaminant. Usually an advanced degree and some experience are required. Must have strong quantitative, statistical, and sampling skills.

NATURAL RESOURCE ECONOMISTS: There will be many new opportunities at all levels. High-level economists with advanced degrees in resource economics develop and justify methods for valuation of injured natural resources. Lower-level economists apply the valuation methods. A background in natural resource management, economics, and environmental regulation is helpful.

ENGINEERING THE ENVIRONMENT

Within the ranks of those actually "fixing" environmental problems, environmental engineers rank high. While engineers spend their careers "righting wrongs," their first, and perhaps most tedious, problem may present itself at the very beginning, when they first consider this field as a vocation. Not only do the uninformed confuse them with railroad people; even among fellow technologists environmental engineers get confused with geologists, chemists, construction engineers, and environmental scientists.

Part of the reason for this confusion is that environmental engineers (that is, those holding that specific college degree) are still a relatively elite group—only a few thousand have actual academic training in environmental engineering. Another difficulty is that many employers, including the EPA, use the term to describe a type of work rather than a type of profession. An "environmental engineer," in common phraseology, is someone who applies technology to environmental problems.

Despite all this, environmental engineering is coming into its own as a highly sought after profession. Enrollments have been climbing at college campuses (many of which have started environmental engineering departments only recently), and job offerings, though not plentiful in these recessionary times, are at least growing in number even as they shrink in certain other engineering disciplines.

"Regardless of the specific engineering degree a job candidate has, the criteria we look most closely for are familiarity with environmental regulations and experience in using engineering skills to address environmental problems," says Jack Taylor, head of the Washington, D.C., engineering department at Environmental Sciences and Engineering. ESE, as it is known in its field, is one of the leading full-service environmental consulting firms. Its engineers, scientists, and business managers handle site investigations and regional surveys of pollution and oversee the remediation work done at contaminated manufacturing plants, waste-disposal sites, and other seriously polluted areas.

HISTORY

Environmental engineering, per se, only came into being after the great wave of environmentalism in the 1960s. Previously, both at a professional and an academic level, environmental engineers were known as "sanitary engineers." Most of them learned the tools of their trade in civil engineering departments across the country.

Sanitary engineering has a grand and glorious history. In the mid-1800s, city governments throughout the eastern United States sought to upgrade their economic power (and protect the lives of their citizens) by funding the construction of irrigation systems that efficiently brought clean, fresh water to the residents and subsequently carried away waste. The critical need for these types of extensive systems was demonstrated during the recurring epidemics of yellow fever, cholera, and other waterborne diseases that devastated urban areas until about 1880. The success of these systems in battling urban disease plagues is memorialized in statues raised in tribute to Benjamin Latrobe, George Waring, and other technological pioneers of sanitation.

By the 1950s, there was a well-established professional organization of sanitary engineers; it changed its name in 1967 to the modern American Academy of Environmental Engineers. Sponsored by eleven other professional organizations concerned with public health and the environment, it offers certification as a "diplomate," a prestigious title for an extensively trained, well-experienced environmental engineer. Included in those sponsoring organizations are the professional societies of the chemical, civil, and mechanical engineers. Thus, in a formal sense, you can obtain a degree in any of those engineering disciplines and, with the right experience in environmental work, call yourself an environmental engineer.

Today, only a limited number of schools offer a degree in environmental engineering specifically, and most of those grant the degree at the master's level or above. *Peterson's Guide to Graduate Programs in Engineering and Applied Science* lists 101 schools that offer civil and environmental engineering degrees; in most cases, the graduate takes a degree in civil engineering, with environmental engineering identified as a specialty. The number of degrees in environmental engineering granted each year, as recorded by the Engineering Workforce Commission (Washington, D.C.), reliably tracks the ups and downs of environmentalism as a social force during the past twenty-odd years: rising during the 1970s, falling during the early 1980s, and then rising again—to new heights—during the past five years. Graduation trends for bachelor's-level civil, chemical, and mechanical engineers are also shown. Being much broader-based disciplines, these do not follow the swings in environmentalism that have been seen in the past two decades. But they are also part of the pool of new graduates that feed into the current professional ranks.

Suzanne Shelley, editor of a new publication, *Environmental Engineering*, looks on the field primarily in terms of the technology being applied. "We define our field as combining the sciences of chemistry, biology, geology, and geophysics with the chemical, civil, and mechanical engineering disciplines," she says. The magazine, currently a supplement to the McGraw-Hill publication *Chemical Engineering*, is being directed toward readers with those backgrounds in the manufacturing industries, government, and consulting groups.

Shelley represents a good example of how technical professionals in the environmental field are adapting their varied backgrounds to their work. She has bachelor's and master's degrees in geology from Colgate University and says that the geological background is a good foundation. "Geology is a fairly interdisciplinary program, involving many different sciences," she notes. "In addition, it provides specific knowledge of soils, rocks, and hydrology that anyone doing environmental engineering needs to have."

ON THE JOB

Environmental engineering is primarily concerned with what the federal government calls "PABCO," pollution abatement and control. This definition implies, as is all too often the case, that pollution is addressed only after it has been generated—after the pollutants go out the smokestack or into a body of water or into a hole in the ground. This type of work has been called, disparagingly, "end-of-pipe" treatment, by which pollution is controlled as it exits the home, car, or factory. Though most forward-thinking environmentalists have realized that the cheapest form of pollution control is usually not to generate it in the first place, the reality is that millions of tons of waste, trash, and sludge spew out of our cities and factories annually. This pollution must be handled on an ongoing basis.

Another buzzword important to environmental engineering is *multimedia*. Whereas the wave of environmental laws passed during the 1970s compartmentalized pollution in the form of contaminated air, water, or solids, the reality in the 1990s is that pollution gets transformed from one type to another as it is processed. The sulfur emitted from a utility smokestack, for example, is turned into a liquid slurry by scrubbing equipment; this slurry is filtered into a semisolid material that is often dumped in a landfill. In most cases, pollution does not disappear; it is simply turned into another substance.

What do environmental engineers do about these streams of pollutants? The simple answer is, they apply technology to reduce pollution from hazardous, toxic forms to safer (if not completely harmless) forms. And whatever residue is left after the processing is done must be properly disposed of, usually in a landfill.

The technology for pollution control involves chemical, civil, and mechanical engineering, primarily. The pollution-control equipment on a modern factory, for example, very often resembles a factory itself. A pollutant is the "raw material"; the environmental engineer sees to it that this feedstream is collected, treated, and disposed of by the safest, most economical means available.

Because its roots are deepest in civil engineering, environmental engineering most often concerns itself with water pollution. A modern wastewater treatment plant (or, for that matter, a water treatment plant that produces potable [drinkable] water) is simply another type of "factory." Dirty water is piped into holding ponds; treated physically, chemically, and biologically; and discharged as a cleansed "product." There are many variations on this theme: treating contaminated groundwater, turning wastewater into potable water, "closed loop" processing in which the same water is used again and again.

Besides a working knowledge of the chemistry, physics, and technical methodology for performing these types of pollution control, the environmental engineer must also have an intimate familiarity with government regulations. Over the years, a massive body of literature has been developed on what is permissible and what is forbidden in controlling pollution. This mountain of documentation changes continually and varies from state to state. Thus, a familiarity with regulations, and how they are interpreted and enforced, is critical to successful environmental engineering.

The employment scene is varied today, and undergoes continual evolution. Many new environmental engineers begin with a job at a city, state, or federal regulatory agency. At the local level, government employment is primarily a matter of reviewing documentation on pollution-control projects and ensuring that regulations are being met. In private industry, many environmental engineers work at engineering consulting companies, which are typically the contractors hired by government or industry to design and build treatment plants or act as project managers during cleanup activities. A smaller number of environmental engineers work for manufacturers themselves, managing treatment works or waste disposal projects, especially in the mining, utility, chemical, and petroleum industries.

Pay scales vary with level of education and type of employer. The lowest-paying are bachelor's-degree jobs with government regulatory agencies—around $28,000, depending on the region of the country. The best-paid are master's degree or doctorate holders working at consulting companies. Starting pay at these organizations is in the range of $30,000–35,000; with several years' experience, and project management skills, the pay can reach the $50,000–60,000 range. Obtaining a professional engineer's license is an added boost.

"There are three things that distinguish the best job candidates," says ESE's Taylor. "First, you have to be a good engineer, with the problem-solving skills to understand, and then resolve, technical issues in control and cleanup projects." Second, he says, is the ability to work with teams, utilizing, in particular, the communication skills necessary to deal with other professionals, such as hydrogeologists, toxicologists, lawyers, or public health workers. And the third might be called, for want of a better term, "strength of character." Environmental engineers are called on to deal with highly stressful situations in which technical issues must be explained to a critical audience—for example, community leaders in the area surrounding a cleanup site. At the same time, the engineer must be able to deal with the sometimes conflicting demands of government regulators, political leaders, and technical experts. These "hot seat" jobs are at the exciting, challenging leading edge of environmental engineering today—where the nagging environmental problems of current and past pollution generation must be dealt with competently.

A GREAT TIME TO BE AN ENVIRONMENTAL ENGINEER

With military base closures now a reality and increasing attention being paid to water quality and availability, environmental engineers will have their pick of projects. Among the many types of activities that environmental engineers are involved in these days, two stand out: designing and running wastewater treatment plants, and cleaning up waste dumps or chemical spills. The latter activity, in turn, divides into two key areas: cleaning up the wastes left by prior industrial activity, and cleaning up special types of wastes, including radioactive materials, for the federal government. These special wastes occur at military bases (which must be restored to some semblance of environmental purity before being converted to other uses) and at nearly all of the several dozen facilities that various U.S. government agencies have developed since World War II to build nuclear weapons.

The significance of environmental work for the military cannot be understated: During the early 1990s, the fastest-growing component of the Pentagon budget, even as the total budget has shrunk, was environmental spending. In nonmilitary cleanups, the primary piece of legislation is the Comprehensive Environmental Response, Compensation, and Liability Act, better known as "CERCLA" or "Superfund," which has allocated $2–3 billion per year since the mid-1980s to study contaminated sites and to begin their cleanup. The total bill for military and civilian cleanup projects has been estimated at anywhere from $200 billion to $600 billion. It will keep environmental and other engineers busy for decades to come.

How do environmental engineers deal with these massive undertakings? Here is a look at what many of them are doing.

WASTEWATER TREATMENT

We generally do not give much thought to water once it goes down the drain. It simply disappears, and on the few occasions when it does not, a call to the plumber usually takes care of the problem. Nevertheless, over the past twenty or so years, the United States has committed hundreds of billions of dollars to cleaning up sewage and runoff from land and city streets. Private industry, a massive user of water as well, has spent billions more. But the problems are not fixed forever; rather, to maintain the lifestyle to which we are accustomed, and to maintain industrial productivity, water is being used more heavily than ever before. In some water-starved regions of the country, cities and factories are now reusing their water again and again rather than letting it drain away. In many cases, that is possible only because modern wastewater treatment plants are able to cleanse water to a higher quality than is commonly available from lakes and streams.

Most of the time, a municipality or industrial complex is not installing wastewater treatment facilities for the first time. More often, existing facilities are being upgraded, expanded, or modernized. The first step is usually a survey of existing water uses, water quality, and projected demand. Is there new housing construction in the area of existing water sources (rivers, lakes)? Are there new potable water demands that will change the availability of water? Do new regulations require different treatment technology?

With a survey in hand, the environmental engineer must meet with city officials, state and federal environmental authorities, and industrial clients to determine future plans. In years past, the federal government was very generous with so-called construction grants to support new projects, but that avenue has been all but closed off. Finding the required funding might necessitate meeting with the financial authorities of local government to raise funds through issuance of municipal bonds.

Assuming that the funding is in hand, the environmental engineer then has a hand in selecting the appropriate measures to be undertaken. This is the technologically exciting element of a project. New methodology, in the form of different types of biological or chemical treatment, must be evaluated. Most municipalities have "primary" and "secondary" wastewater treatment, which involve the use of settling basins and sludge processes where sewage is cleansed of heavy solids. In many regions, tertiary treatment is now required, which may necessitate anything from ultraviolet radiation to carbon filtration to cleaning the water thoroughly.

Another technical issue that wastewater treatment managers are currently dealing with is the disposal of sludge—the residue left over after water has been treated. Until recently, many coastal regions were dumping this at sea, but that practice is now banned. Sludge can be disposed of on land—but at a price. Environmental engineers must weigh the cost of land disposal against alternative treatment such as sludge drying (to reduce its volume) or "anaerobic processing," which converts much of the sludge to methane gas.

With all the plans in place, a "request for proposals" is issued, opening the bidding to private contractors to design and build the treatment facility. This phase of a project is called "contract management" and generally has very specialized requirements to ensure that competing bids are evaluated fairly and that the technical demands of the project are met. Contract management follows the project through to construction and commissioning.

As these situations demonstrate, the environmental engineer must be adept at diverse skills, ranging from negotiating with government officials to deciding highly technological issues.

HEALING THE LAND

By now, the problems of unregulated disposal of hazardous wastes across the country have been well documented. The U.S. Environmental Protection Agency has a list of some twelve thousand sites where wastes have been dumped in an environmentally harmful manner. But the problem is much more insidious than a few corporate bad actors. A century of gas station construction—built, abandoned, and forgotten—has left a legacy of leaking underground storage tanks that frequently allow toxic fuels to seep into underlying water tables. Even homeowners are finding that old oil tanks on their property can bring down the wrath of environmental officials.

As bad as they are, these problems pale in comparison to those of the military and nuclear dump sites. There, wastes ranging from weaponry (including unexploded shells at firing ranges) to the by-products of chemical warfare agents have been sloppily disposed of. The network of nuclear processing facilities that built warheads and supplied nuclear fuel for power production is tarred with radioactive agents, heavy metals, toxic solvents, and a witch's brew of other problem wastes. Some of the underground storage tanks holding radioactive liquids at the Hanford site in the state of Washington are chemically active—periodically, they "burp" a bubble of hydrogen gas, itself an explosively flammable material.

EPA has developed a dauntingly rigorous procedure for investigating and evaluating such wastes. The first step (assuming that the dump is relatively stable and not in need of emergency treatment to eliminate an imminent threat) is a remedial investigation and/or field survey—a fancy name for a thorough study of the geology and chemistry of the site. Usually, this activity requires a series of wells to be drilled at spaced intervals throughout the site. Core samples are taken by an EPA-specified procedure and then analyzed in a laboratory for contamination. Calculations are made of the materials present and their concentration, the hydrogeology of the site—the direction in which groundwater moves, the type of soil present, and the potential toxic effects of the contamination.

In recent years, the technical work often stopped at this point, because another part of the survey is to identify, if possible, whose wastes are present. If such a determination can be made, the lawyers go into action, seeking funds from the users of the site. Chemical companies and "culprits" generally countersue, often claiming that the amount of their wastes present is being overestimated. The final settlement can take years to negotiate.

Whether or not the original polluters are identified, at some point an actual cleanup project begins. A great variety of technologies have been developed for this task, and more are under development all the time, because most of them are expensive. The soil can be washed with detergents or solvents, then reinterred. It can be incinerated, driving off or destroying toxic organics. An especially popular new method is the use of microbial colonies, which can be "trained" to attack and digest certain types of organic chemicals. Such "bioremediation," as it is called, offers lower cost, but sometimes is too slow to be the preferred alternative.

Once a suitable type of technology has been identified, environmental engineers go to work in designing the size and nature of the cleanup needed. Sometimes, it is not one technique that is used but an entire series. It is all but impossible to restore a site to its original character, but most cleanup projects are successful enough that no further contamination around the site occurs.

Throughout the 1980s, billions of dollars were expended on site investigations. This has led to some criticism of the Superfund program, because the investigation, in and of itself, does little to improve the environment. According to industry sources, however, the 1990s will be characterized by the actual cleanup of sites, cleanup that could not be undertaken without extensive site investigations to determine the best course of action.

Environmental engineering in this context is partly an exercise in developing and using new technology and partly a legalistic process of interpreting highly complex regulatory procedures for analyzing and then cleaning up a site. Training in the appropriate laws and regulations has already become part of the curriculum of many environmental engineers. As with wastewater treatment projects, good skills in communications, reporting, and negotiating are desirable.

Though there are a finite number of Superfund dump sites, there is every indication that cleanup projects will be an ever-present issue for environmental engineers. In both wastewater treatment and land cleanup, the ultimate concern is developing better ways of handling our wastes and creating a society that is more in harmony with the needs of nature.

BALANCING THE SCALES: CAREERS IN ENVIRONMENTAL LAW

In a democratic society, law both regulates conduct and reflects certain policy choices. Environmental law reflects society's choices with respect to the land, water, air, and ecosystems that surround us. In seeking to protect the environment, environmental law affects all aspects of human activity. The cars we drive, the way we work, the use of our land, and the products we use are all influenced by the policy choices known as environmental law.

Because environmental law exerts such a profound influence on human activity, there is a great need for individuals trained in navigating the complex maze of regulations, statutes, and court decisions that form the basis of environmental law. Though many of these individuals are lawyers, environmental professionals in all fields increasingly need some familiarity with environmental law to do their jobs. Routinely, position descriptions require "knowledge of environmental laws, regulations, and legal issues" as a prerequisite for employment.

THE WORK

Lawyers and others trained in environmental law find themselves in a variety of positions and contexts. Generally, the work may be divided into five areas:

- ⇢ Compliance counseling to assure that the requirements of environmental laws are met
- ⇢ Enforcement by government agencies of environmental laws
- ⇢ Preparation of documents in transactions having an environmental impact
- ⇢ Litigation of disputes involving environmental issues
- ⇢ Influencing of environmental policy through regulations or legislation

In each of these areas, a specialized knowledge of environmental law is required.

Compliance Counseling

Compliance counseling is designed to prevent environmental problems before they occur. It may take the form of reviewing clients' business practices to determine if they comply with new regulations or statutes. It also includes assisting clients in complying with environmental regulations in new projects. For example, a business interested in building a new manufacturing plant must comply with environmental regulations on federal, state, and local levels. Often this means preparing permit applications, attending zoning hearings, or compiling environmental impact statements.

Enforcement Work

Enforcement work is performed by attorneys at all levels of government. Environmental laws are enforced through civil damages, administrative fines, and, in some instances, criminal proceedings. The attorney reviews the data prepared by staff investigators, determines if there was a violation, and ascertains the appropriate procedure to remedy the violation. Often enforcement actions lead to litigation in the form of court trials or contested administrative hearings.

Transactional Work

Transactional environmental work focuses on negotiating and preparing documents for economic transactions that have environmental effects. This area includes negotiating and preparing documents for scenic easements or preserving development rights. In recent years, some of this work has been done internationally in conjunction with debt-for-land swaps, which have enabled preservation of rain forest lands in Central and South America. Under the revisions of the Clean Air Act, new transactions will be permitted in which businesses will be able to trade "pollution rights" if their emissions are less than permitted by law. This area will occupy an increasing share of environmental lawyers' attention.

Litigation

By far the most visible area of legal environmental work is litigation. We are all familiar with the litigation that has sought to protect endangered species such as the spotted owl or the snail darter. Attorneys working in this area review projects based on their impact on the environment and litigate on behalf of environmental groups in an effort to halt destructive projects. Among the best-known environmental litigators are the Natural Resource Defense Council and the Sierra Club Legal Defense Fund.

Environmental Policy

Many people trained in environmental law work to influence environmental policy. This includes researching proposed legislation and regulations, lobbying government officials, and organizing environmental campaigns. Additionally, many industry and trade organizations actively work to influence environmental law and policy and to educate their members on the impact of new environmental regulations. Policy work often requires interaction with scientists, government officials, and community organizers.

PREPARING FOR AN ENVIRONMENTAL LAW CAREER

Lawyers are typically trained as generalists rather than specialists. That is to say, they are trained to be *lawyers*, not criminal lawyers or tax lawyers. Because of this, most environmental lawyers and environmental professionals who work with environmental law have learned it in the field. Because of the increasing complexity of environmental regulation, however, a growing number of law schools are offering specialized programs, courses, and degrees in environmental law.

Law school requires a strong academic record but not a specific academic major. Many environmental lawyers have found it helpful, due to the complex technical issues addressed in environmental regulations, to have a background in physical sciences or engineering, while some students come from liberal arts, business, or environmental science backgrounds.

Because legal education occurs only at the professional school level, students must generally take the Law School Admissions Test for entry into law programs. Because environmental law is a specialized area, some law schools offer extensive course work only at a master of laws level (L.L.M.). A number have combined the law degree with a separate master's degree in joint degree programs, or offer certificate programs in environmental law. This latter approach enables the student to receive extensive environmental law course work in three years rather than the four years required for the L.L.M.

Many environmental professionals take courses in these master's programs to obtain the legal knowledge needed in their careers, without needing to be able to practice law. Environmental engineers, for example, may take courses to learn the regulatory structure of the Superfund legislation (CERCLA) to better advise their clients on cleanup requirements. Environmental activists might take environmental law courses to increase their effectiveness in lobbying legislators on environmental issues.

Environmental law programs generally offer courses in three basic areas:

- ↬ Legal foundation courses, such as legal research and writing, that are designed to expose the student to the sources of environmental law and develop skills in locating the applicable regulations

- ↬ Specialized environmental courses that focus on particular environmental laws or particular topic areas, such as historic preservation or coastal zone management

- ↬ Administrative law courses, focusing on the regulatory process, how environmental regulations are written and enforced

Because environmental issues are increasingly global in scope, it is likely that environmental programs will adopt some international component. These programs could include courses on environmental law in particular countries or issues that transcend national boundaries, such as global warming and ozone depletion.

INTERNSHIPS AND SUMMER POSITIONS

Law students are encouraged to use their summers to gain experience putting their legal skills to use. Positions are available on both a paid and volunteer basis with a variety of organizations. In recent years, students have worked with employers such as the U.S. Department of Justice, Environmental Protection Agency (EPA), the National Wildlife Federation, the United Nations Environmental Programme, major corporations, and environmental departments of law firms. Internships are often available with environmental organizations on the grassroots level. These experiences provide students with firsthand knowledge of the practice of environmental law as well as provide an opportunity to network for permanent positions upon graduation.

Some law schools provide students with the opportunity to gain experience under direct supervision of faculty instructors. These "clinical" programs function as law offices in which students represent actual clients in real cases.

CAREER OPPORTUNITIES

Environmental lawyers and environmental professionals with legal training work for a diverse group of clients in both traditional and nontraditional settings. These opportunities may be found in the public, private, and nonprofit sectors. As environmental problems become recognized as international in nature, there is a need for individuals with legal training to address these problems on a global scale. The positions involve working on significant environmental problems, issues, and policies, and making a contribution to the resolution of environmental disputes.

Private Sector Positions

As you might expect, most environmental lawyers work for law firms. Large law firms have extensive environmental law departments that may employ as many as fifty attorneys to handle environmental cases for their clients. In this setting, lawyers typically advise business interests on how to comply with environmental laws or represent them in disputes about environmental violations. Nonlawyers with environmental law training are also employed by large firms as paralegals or environmental analysts to assist clients in processing permits for development projects. Smaller firms often specialize their practice exclusively to environmental law and may represent corporations or municipalities in environmental and land-use cases.

Corporations make extensive use of individuals trained in environmental law. Larger corporations have environmental lawyers as part of their in-house legal departments. A large number of both lawyers and nonlawyers are employed to assure compliance with environmental regulations. Because of the technical nature of many corporations' business, those working in environmental compliance often also have degrees in engineering or chemistry.

Insurance companies also use environmental lawyers and nonlawyers to process and evaluate claims for environmental damage. In this setting, the environmental lawyer or claims manager must determine if the company is liable for the claim, investigate the cause of damage, estimate the cost to repair the damage, and finally negotiate a settlement of the claim.

Consulting firms make up one of the fastest growing markets for environmental professionals with legal training. Consulting organizations work for both business and government to assist them in developing environmental standards or in complying with environmental regulations. These firms may be used by smaller organizations that lack the needed expertise on staff or may work on special projects with environmental impacts.

A nontraditional employer of environmental professionals with legal training is publishing companies. These companies specialize in providing information on environmental laws and regulations to managers in business and government. Those employed in these positions monitor the status of proposed environmental regulations and the outcomes of court cases to keep their subscribers updated with current information. Trade organizations, such as the National Manufacturers Association, or the American Chemical Society, perform a similar function for their members. However, in addition to educating members with regard to environmental laws and regulations, these groups also actively seek to influence environmental law and policy in favor of their members' interests.

Public Sector Jobs

Because government functions by law and regulation, it is no surprise that the government hires a large number of lawyers in the environmental area. This is particularly true because our environmental laws are administered by a number of government agencies. The EPA, Department of Interior, Department of Agriculture, Army Corps of Engineers, National Oceanographic and Atmospheric Administration, Food and Drug Administration, Department of Transportation, and Department of Labor all play some role in developing and enforcing environmental law and policy at the federal level. A similar situation exists on the state level. City and county attorneys are also called upon to deal with environmental issues as well.

Nonlawyers with legal training working in the public sector have a variety of job titles, but all involve applying environmental law in particular circumstances. For the EPA, an environmental specialist might work with federal, state, and local governments to develop policy or regulations under federal statutes. At the state level, such an individual might be called an environmental program manager and may process permit applications, interact with federal and state agencies, provide expert testimony, and develop regulations under state coastal zone legislation. Locally, the position might be termed an environmental supervisor, with responsibility for compliance monitoring and enforcement of regulations related to wastewater treatment or solid waste disposal. These positions are often on the front line of enforcement of environmental law and development of new environmental regulations.

Nonprofit Positions

Nonprofit organizations actively seek to influence and, in some cases, enforce environmental law and policy. Training in environmental law is crucial to the ability of these organizations to affect policy decisions, organize citizen campaigns, and disseminate information regarding environmental issues. It is obviously difficult to advocate for a change in environmental law or policy without an understanding of the current state of the law or policy.

Lawyers in nonprofit environmental organizations seek to influence policy in a variety of ways. Many do it through monitoring legislation and lobbying decision makers for changes in statutes or regulations. Others might spend large amounts of time researching and writing on environmental issues in the form of comments to proposed agency actions. Organizations that pursue these strategies on the national level include the National Wildlife Federation, the National Environmental Law Center, The Nature Conservancy, and the Audubon Society. Similar organizations work on state and regional levels to influence environmental policy.

Lawyers also engage in litigation on behalf of environmental organizations and community groups. These lawsuits seek to challenge actions taken by governmental or private organizations that are viewed as detrimental to the environment. Lawyers in these organizations gather data, research the law, argue cases in court, and negotiate to obtain positive environmental results. Among groups performing environmental litigation are the Sierra Club Legal Defense Fund, Natural Resources Defense Council, Environmental Defense Fund, and Defenders of Wildlife. Recently, some environmental litigation groups have been formed to protect the rights of unrepresented citizens in environmental decisions. These groups have challenged the use of pesticides around migrant farm workers or the location of incinerators in minority communities.

Some environmental groups work to conserve particular natural resources for future generations. Lawyers working for these organizations work on structuring transactions to assure the preservation of the resources. Among these organizations are The Nature Conservancy, the National Trust for Historic Preservation, the Trust for Public Land, and The Wilderness Society.

Nonlawyers with some training in environmental law may work for any of these types of organizations in a number of capacities. Many serve as organizers around a particular issue, such as toxic pollution, clean water, or resource preservation. Positions such as program director, policy analyst, or environmental associate often call for a background in environmental law and policy. Important roles in all nonprofit organizations are fund-raising and public relations. These positions may also be filled with nonlawyer advocates or environmental professionals.

SALARIES AND FUTURE GROWTH

Environmental law is an emerging field, and the outlook for positions in all sectors remains strong. Increased environmental awareness and a more active government posture toward protection of the environment will tend to increase demand for environmental professionals with at least some legal training. Environmental professionals in such diverse fields as environmental engineering, industrial hygiene, and planning will increasingly encounter legal issues as part of their everyday work. This in turn will demand further efforts to educate environmental professionals in environmental law.

Salaries for positions in the field of environmental law depend on both the nature of the position and the nature of the employer. For lawyers, entry-level positions with large law firms and corporations can range from $40,000 to more than $60,000 per year. Government entry-level positions generally range from $27,000 to $35,000, depending on the level of government. National nonprofit organizations rarely hire at the entry level, but experienced attorneys with nonprofit organizations generally earn between $30,000 and $40,000. At smaller nonprofits, which do hire at the entry level, salaries are often as low as $20,000 per year.

Salaries of professionals with legal training are largely dependent upon prior background and experience. An experienced engineer with extensive training in Superfund regulations can earn in excess of $40,000 with corporations and consulting firms. Government environmental specialist positions will often base salaries on other additional factors (e.g., science background) but generally range from $25,000 to $30,000 at entry level. Those working in nonprofit environmental organizations tend to earn between $17,000 and $25,000 at entry level.

Environmental law as a career offers the opportunity to remedy past environmental problems, to regulate current activities to protect the environment, and to plan for future policies that will result in sustainable growth. Both lawyers and nonlawyers with legal training will participate in aspects of these opportunities in a variety of settings.

Chapter Nine

THE GREEN WORKPLACE: CHANGE AND DIVERSITY

THE DIVERSITY DEBATE

Diversity is the latest buzzword of the 1990s, but an old Chinese proverb best describes the need for talking about workforce diversity: "If we do not change the direction we are going, we end up where we are headed." The question now is, How smooth will we make the journey? The "diversity wave" is more than trendiness. Diversity is an issue continually gaining importance in people's lives, in politics, science, technology, and the environment. Nowhere will there be more of a need to negotiate diversity and unity than between the environmental and minority communities.

Some organizations are leading the way by establishing task forces, sponsoring conferences, and implementing comprehensive programs for outreach, recruitment, and training. Others are on the sidelines waiting to see what will happen; others just do not understand what this is all about.

This increased concern for diversity is justified by forecasts in the Department of Labor's *Workforce 2000*. This report predicts that by the twenty-first century, 75 percent of new employees in the labor pool will be women, minorities, and immigrants. White males make up only 47 percent of today's workforce. In addition, the 1990 census has recorded dramatic population growth within our minority communities since 1980. The Asian and Pacific Islander community alone has grown 107.8 percent since 1980, while the Hispanic community has increased 53 percent. (The Native American population has increased 13.2 percent, and the black community 6 percent.)

America is changing in other ways. The Department of Labor's demographics forecast an older workforce with fewer young people entering the marketplace. In addition, the 1990 census figures show a dramatic dispersion of minority groups into all areas of the country in the past decade. Although minority populations are still predominant in our urban communities, their numbers are increasing in suburban and rural areas. America is also becoming more economically polarized. The rich are richer, and the poor are poorer.

The United States is more culturally diverse than at any other time in its history. In a recent article in the *Multicultural Network Newsletter*, Shakura A. Sabur, the director of the American Society for Training and Development's Multicultural Network, stated, "The workforce is changing. The workplace must also change. Organizations must learn to replace the institutional monocultural norms with the institutionalization of multicultural values and norms."

In a recent interview, former Forest Service chief Dale Robertson voiced his views, "Diversification without management is chaos. We are not talking about hiring people and trying to mold them into Euro-American values. They should bring their own cultures and values with them and integrate them into our work." The challenge is to create "unity in diversity," a phrase emphasized by Dr. Chester M. Pierce, a professor at Harvard University's Schools of Psychiatry and Education and a Student Conservation Association director. "Infusion of different inputs and viewpoints promotes unity and strength for everyone," he notes.

WHY DIVERSITY IN THE CONSERVATION COMMUNITY?

Social and environmental movements go hand in hand. Poor and minority communities are the chief victims of pollution and other environmental hazards. High levels of lead, mercury, and toxic poisoning are much more prevalent among poor and minority populations. Meanwhile, environmental groups are realizing the need to educate and involve a broader-based constituency to promote protection of natural resources.

Terry Ow-Wing, a member of the Sierra Club's Ethnic and Cultural Diversity Task Force, says, "In the environmental movement, we have all learned that the world is small and finite. If we in the movement forget minorities who live in the urban environment, a critical part of solving the greater problems will be missing." Ow-Wing has made a commitment to begin a process of mutual education. Although the task force is new, she sees that it can send a clear directive that all people must work together to protect the environment, and that neglect of any sector of the population will serve to undermine a true sense of environmental equality.

The environmental field also is seen as a source of opportunities for minorities. Employment in environment and energy organizations will continue to grow (though perhaps at a slower rate than over the past ten years) for the foreseeable future. All of us must be committed to increasing the diversity of the environmental community by helping people of color identify and achieve professional positions with the movement.

THE CHALLENGES OF DIVERSITY

In recent years, many conservation organizations have been criticized publicly for the lack of minority representation among their staffs and boards. Ask conservation leaders, and many will say that it is not that they have not tried but that there are not enough minorities who are interested or prepared for careers in natural resource management. Ask minorities, and many will say that the movement is driven by those who resist change. Whatever the reasons, the fact remains that the conservation profession is a predominantly white male workforce. Regardless of the viewpoint, the need to diversify is clear.

The Conservation Leadership Project (CLP) conducted by the Conservation Fund in 1989 surveyed more than five hundred conservation leaders. The project noted that "virtually none of the mainstream groups . . . works effectively with, or tries to include people of color, the rural poor, and the disenfranchised." Among the six distinct strategies to improve organizations' effectiveness at all levels of the conservation movement, the CLP report noted the need for participation by minorities and other nontraditional groups.

The challenges to diversify the conservation profession are many. Historically, minorities have been underrepresented in environmental organizations. Notably, women and minorities have been rarely placed in upper management positions. Second, the priorities of many organizations do not actively address or value the issues and environmental concerns of the minority community. Third, minorities have often depended economically on the very organizations and corporations that environmental organizations seek to change. Finally, the cultures of many of these organizations are not prepared to embrace, celebrate, and manage the strength and productivity that a diverse workforce can offer.

But for many others, challenges are opportunities. Robin White, a supervisory park ranger at Indiana Dunes National Lakeshore in northwestern Indiana, has spent much of her career forming partnerships between the urban and minority communities and the park. "The barriers are difficult to overcome and will vary in their effects on different groups," she notes, "but the establishment has the power to evoke change. The work of change is very rewarding."

JOINING HANDS FOR SOCIAL AND ENVIRONMENTAL CHANGE

The ultimate goal for the conservation movement and minorities is to come together for global understanding and protection of our environment. We not only have to increase the numbers of women and minorities in the conservation profession but also increase partnerships with minority communities that serve to promote mutual trust and understanding.

When Matthew Mukash, a Cree Indian, became aware of a hydroelectric dam proposed on Canada's Great Whale River, he knew he had to do something. The dam would not only change his people's land and wipe out their livelihood but have a devastating impact on wildlife. The first dams of the James Bay Project were completed by the government-run Hydro-Quebec utility company in 1985 and flooded more than 4,400 square miles of subarctic wilderness in northern Canada. Mukash says that more than ten thousand caribou were drowned, and the habitat of millions of migratory birds, brook trout, moose, and muskrat was destroyed. Mukash began a campaign to bring together the Grand Council of the Crees and conservation groups. Mukash is excited about the diversity and large number of groups, including the Sierra Club, that have joined together against the project. "This was really the beginning of our working together," he says.

Their efforts seem to be paying off: The second phase of the project has been postponed, and Mukash is still hopeful that the three court cases pending against Hydro-Quebec and the Canadian government will push the utility into filing environmental impact statements for the project. This is an example of minority and environment organizations joining not only in protection of the environment but also in preserving the heritage and history of Native Americans. We forget that conservation and preservation of natural areas often means preservation of cultural resources, particularly within many of our federal parks.

Consider Robert Stanton: Stanton is a groundbreaker. He became the National Park Service's first black superintendent and has gone on to a distinguished career as regional director of the National Capital Region, which includes thirty-six federal parks and historical sites. The region attracts more than 40 million visitors annually to popular locations, ranging from the White House to the Vietnam Veterans and Lincoln Memorials, to the Frederick Douglass Historic Site. Stanton believes that a direct relationship between conservation organizations and the leadership of civic organizations, public schools, and higher education has benefits for all. "The real opportunity for the minority community is to understand what is available within the National Park System," says Stanton. "One of the most rewarding opportunities for me has been to participate in making my history and my roots part of America's heritage." Recently Stanton testified before Congress to include the Mary McLeod Bethune Council House and Archives as a national historic site in the system. "When I started in the National Park Service, there were only three sites dedicated to black history; today there are nine," he notes.

Private organizations that recognize that preservation of natural and cultural resources go hand in hand include the Conservation Fund, which is currently seeking to preserve Port Hudson Battlefield in Louisiana. Port Hudson has been designated by the secretary of the interior as one of the twenty-five priority Civil War battlefields. Why this terrain? The battlefield is the site of the first major assault of black soldiers in the war on May 27, 1863, a place where Americans can learn about the vital role of black soldiers in history.

Charles Jordan, director of parks and recreation for the City of Portland, Oregon, states, "There has never been a better time in the history of the environmental movement and the struggles of minorities to come to the table with contributions beneficial to both." Jordan, former chair of SCA's Conservation Career Development Program (CCDP) Advisory Committee, was a key member of the Conservation Fund's Leadership Study Task Force to look at minority participation in the conservation profession. In addition, he served on the President's Commission on Americans Outdoors.

A theme Jordan tries to include in all his environmental work is that the social problems of the inner city and the ecological problems of the planet have some common solutions. He advocates the social benefits of recreation in natural settings. "Leisure provides important social benefits such as conflict resolution and self-esteem building," he explains.

Surely now is the best time ever for the environmental and minority communities to join hands. The increase in population brings more diversity. Unfortunately, it also tends to bring more pollution. The once-clear vista at Grand Canyon National Park is marred with the smog of urban centers far away. Everyone has to be involved in the environmental movement, or we all lose.

REACHING OUT

How do we attract a more diverse group to conservation? The approaches and programs need to be direct and sensitive, stresses Flip Hagood, chief of employee development for the National Park Service. He notes, "We need to start more outreach work. It is up to the professional to make the changes." Hagood believes focusing on the younger generation will generate the highest returns. In fact, the Park Service has set up programs to attract minority college students as seasonal rangers. He adds, "I believe in direct one-to-one contact: We as professionals need to send minorities who are actual practitioners to the universities to begin attracting other minorities into the field. We should not send programmers, specialists, or equal opportunity leaders."

Al Worthington, a personnel officer with the Forest Service, stresses that "young minorities need to be prepared and committed and need to look for things on the cutting edge. They also need math and sciences. They need to understand what a conservation career involves." Asked what "worldly advice" he would give to minority students, Hagood says, "Explore the possibilities that this profession has to offer. This profession directly links us to our historical and cultural heritage. We soon realize that we need to save our world and our earth."

DIVERSITY AT WORK IN THE CONSERVATION PROFESSION

Despite the need for greater diversity, it is already at work in the conservation field and growing stronger every day. Over the past two decades many individuals and organizations have contributed significantly to diversity in the conservation profession. Some have been trail blazers in breaking down stereotypes and setting the example for others. A small sample of the individuals and organizations working for change follows.

PEOPLE

Vickie Glover

Vickie Glover is a program management specialist at the George Washington National Forest in Virginia. What makes Glover special is not merely her outreach work, but the approach she shares: "People are people; we need to look at the individuals and break down stereotypes. We will be better off when we see how similar we are."

In addition to her responsibilities for overseeing all George Washington human resource programs, Glover is the Native American program coordinator. As a Native American herself, she is a role model for younger Native Americans considering conservation careers.

Vickie has begun a cooperative agreement with a local business school to allow students to earn credit for time spent working at the forest. The students do not receive pay, but the experience and exposure they receive are very valuable. In turn, managers have an excellent opportunity to determine the potential of the students for careers in the Forest Service.

Jose A. Zambrana

Following two years of agricultural studies, Jose Zambrana began his career as a forest technician in his native Puerto Rico at Caribbean National Forest. Recognizing his talent and desire, the Forest Service assisted Zambrana in completing his education. In 1969 he received his bachelor of science in Forestry from North Carolina State University.

Soon after getting his degree, Zambrana moved into a professional series, and his career took off. He worked at Ocala National Forest in Florida and is now district ranger at Uwharrie National Forest in Troy, North Carolina. "The most challenging part of my job," he says, "is mixing human and natural resources to provide services and keep the people happy."

Zambrana is busier than ever; his forest was granted $2.83 million for land acquisition and recreation development. Despite the busy pace, Zambrana makes time to counsel others, noting that "the key to success is education—be ready to meet the opportunity. Nobody can take away what you have learned."

Zambrana hopes to assist the Forest Service's international programs, stressing that "there is a need to share our skills with other countries and assist them to manage their forests."

Vivian Li

To her knowledge, in 1985 Vivian Li became "the first person of color on the board of directors of a national environmental organization," the influential Sierra Club. (She was glad to learn that George Patterson had earlier become a member of the SCA board of directors.) It was a historic moment for Li, considering that she assumed the position through election by the large, activist Sierra Club membership. Vivian Li is used to being a trailblazer. After she graduated from Barnard College in environmental studies in 1975, she became one of the few non–African Americans on the staff of the planning department in Newark, New Jersey.

In 1978, working with the National Urban League in New York, she coordinated the Citycare Conference, which brought together eight hundred environmental and civil rights activists for the first time. After that, she worked closely with the Sierra Club urban environment committee and chaired its air quality committee, where she helped devise the Club's policy on hazardous wastes, keenly aware that "people of color are more adversely affected by incinerators and toxic wastes. You know darned well they won't site a facility like that in Scarsdale. Those communities that tend to be poor are most vulnerable, and they tend to be communities of color." In 1989, Li helped the Sierra Club set up its Ethnic and Diversity Task Force, which aims to increase the numbers of people of color on the Sierra Club staff as well as diversify Club membership by ethnic group, race, class, and gender and cooperate with other groups in diversifying the environmental movement.

Deejohn Ferris

Deejohn Ferris has served as the director of the environmental quality division of the National Wildlife Federation. She began her career in the environmental field upon graduation from Georgetown University Law Center in 1970, first working at the EPA's Office of Litigation. During her eight years with EPA, Ferris served as director of the special litigation division and dealt with issues ranging from the use of pesticides to regulatory policy.

The fact that many environmental policies are new and are continually changing and growing is what attracted Ferris to the field. "The environmental field has been and remains a fairly innovative area of the law. There is great latitude for innovative thinking and the development of creative solutions to complex problems," she says.

Commenting on ethnic diversity in conservation, Ferris states that there are many minorities involved in and concerned about a wide variety of environmental issues, with different groups defining specific environmental concerns. "Lead poisoning and clean drinking water are seen as environmental issues by many people but are often referred to as health issues by minorities," she says.

"There are groups of environmental professionals from ethnic minorities that can serve as guides to individuals who are seeking to learn more about the environmental field," says Ferris. Two groups she notes are the Center for Commerce, Environment, and Energy and the African-American Environmentalist Association. "We all need to be involved in issues that affect our daily lives," she states.

PLACES

Native American Fish and Wildlife Society

The Native American Fish and Wildlife Society recognizes that Native Americans as a group have always demonstrated environmentally sensitive practices toward this earth's precious resources. To further the tradition, NAFWS supports tribes and tribal leadership in their self-determined march toward a secure natural resources future.

Since its establishment in the early 1980s, the NAFWS has focused its efforts on a wide variety of national endeavors that promote Native American natural resource management. The organization's services range from technical assistance (helping with grant or proposal writing, serving as a clearinghouse for information and a contact network, and acting as liaison between tribes and other decision-making entities) to educational assistance (providing training for tribal members in fish and wildlife management, scholarship opportunities for tribal youth, community outreach to reinforce natural resource and environmental concerns, and instruction geared toward encouraging interest in professional natural resource career opportunities).

Indiana Dunes National Lakeshore

Indiana Dunes National Lakeshore may be the only national park with three permanent full-time interpretive positions devoted to outreach work, a program initiated in the early 1970s. The park, located in an ethnically diverse region of northwest Indiana, forty-five minutes from Chicago, has coordinators for bilingual and special populations and urban outreach programs. In addition, the park has one of the largest environmental education programs in the National Park System.

Each outreach coordinator is responsible for working with a target population in the surrounding communities. The goal: to take the park to the people. Supervisory Park Ranger Robin White says, "As the park service establishes partnerships with various agencies and organizations, a commitment to challenge in environmental education, prevention, and intervention will outlast the inevitable changes of the park administration." This approach offers services that meet not only the needs of the park but of the surrounding communities as well.

George Washington National Forest

George Washington National Forest in Virginia holds a career camp each summer at which students of various backgrounds gather for an introduction to careers in natural resources. Students from a six-state area listen to speakers and complete field assignments to give them a taste of conservation work.

Three days of the camp are held at Virginia Polytechnic Institute and State University, where professors introduce the academic training and knowledge needed for natural resources careers. The minorities and urban students who attend get an opportunity to break down their own stereotypes about the outdoors and others. "We're learning a lot about the environment and making good friends from so many different areas," reports Jasmine Murray, a participant from Washington, D.C.

CONSERVATION CAREER DEVELOPMENT PROGRAM

Founded by the Student Conservation Association in 1990, the Conservation Career Development Program (CCDP) provides necessary fellowships and conservation leadership training primarily for minorities and women. This year-round program serves high school and college participants seeking field experience, training, mentoring, and higher education in the conservation fields.

The CCDP is designed to recruit and train participants for conservation careers in forestry, wildlife, and natural resource management; park and recreation administration; engineering; planning; archaeology; and history.

In addition to field experience in national parks, wildlife refuges, and forests, the program provides extensive tracking, mentoring, and other career services for CCDP participants, including guiding them to appropriate resources, such as academic and vocational training programs. It also assists qualified participants in securing permanent positions with conservation agencies.

The CCDP also provides diversity training for participants and conservation agencies in order to facilitate the entrance and retention of minorities in the conservation workforce. The CCDP also provides career fairs, educational assessments and placement, a recruiting service, and a speakers bureau.

DIRECTORY OF ORGANIZATIONS OFFERING GREEN JOBS

AAA Engineering & Drafting Inc.
 1865 S Main, Ste 12, Salt Lake
 City, UT 84115
ABB Environmental Systems
 Centerpoint Plaza, Knoxville,
 TN 37932, (713) 931-4400
Access Outdoors, Inc.
 950 Euclid Ave, Ste 303, Miami
 Beach, FL 33139, (800) 673-
 0480
Acid Rain Foundation, Inc.
 1410 Varsity Dr, Raleigh, NC
 27606
*Acid Rain Information
Clearinghouse*
 33 South Washington St,
 Rochester, NY 14701
Accu-Labs Research, Inc.
 4663 Table Mountain Dr,
 Golden, CO 80403
Acres International Corporation
 140 John James Audubon
 Park, Amherst, NY 14228
ACZ Laboratories
 30400 Downhill Dr,
 Steamboat Springs, CO 80487

AD+SOIL, Inc.
 210 Gale Lane, Kennett Square,
 PA 19348
Adirondack Environmental Services
 314 North Pearl St, Albany, NY
 12207
Adirondack Mountain Club
 Box 867, Lake Placid, NY
 12746
*Adirondack Nature Conservancy—
Land Trust*
 Box 188, Elizabeth, NY 12932
*Adirondack Outdoor Education
Center*
 Education Center, Pilot Knob,
 NY 12844
Advanced Systems Technology
 3490 Piedmont Rd, Atlanta,
 GA 30305
Adventurers' Company
 PO Box 2290, Nevada City, CA
 95959
Adventures Rolling X-Country
 2269 Chestnut St, Ste 119, San
 Francisco, CA 94123
African Wildlife Foundation
 1717 Massachusetts Ave NW,
 Washington, DC 20036

Agriculture Research Service
 1700 South West 23rd Dr, PO
 Box 14565, Gainesville, FL
 32604
*Air and Waste Management
Association*
 PO Box 2861, Pittsburgh, PA
 15230, (412) 232-3444
Air Pollution Control Federation
 Air Pollution Control
 Association, PO Box 2861,
 Pittsburgh, PA 15230, (412)
 232-3444
*Air Resources Information
Clearinghouse*
 RD 1, Box 185, Valley Falls, NY
 12185, (518) 753-7838
*Alabama Environmental
Management Department*
 1751 Dickenson Dr,
 Montgomery, AL 36130
Alaska Center for the Environment
 519 West 8th St, Ste 201,
 Anchorage, AK 99501
Alaska Conservation Foundation
 Board of Directors, 430 W 7th
 Ave, Ste 215, Anchorage, AK
 99501

Alaska Environmental Conservation Department
  410 Willoughby Ave, Juneau, AK 99801

Alaska Foundation
  Hulbert Outdoor Center, RR 1, Box 91A, Fairlee, VT 05045

Alaska Natural Heritage Program
  707 A St, Ste 208, Anchorage, AK 99501

Alaska Observers, Inc.
  130 Nickerson St, Ste 206, Seattle, WA 98109, (206) 283-7310

Alaska Division of Parks & Outdoor Recreation
  400 Willoughby, 3rd fl, Juneau, AK 99801, (907) 465-4563

Alaska State Parks
  Dept of Natural Resources, PO Box 10-7001, Anchorage, AK 99510

Alaska Wildland Adventure
  Box 389, Girdwood, AK 99587, (907) 783-2928

Alaska Wildlife Alliance
  Box 202022, Anchorage, AK 99520

Alabama Natural Heritage Program
  State Lands Div, 64 N Union St, Rm 752, Montgomery, AL 36130

Alcoa, Inc.
  PO Box 300, Bauxite, AR 72011

Allagash Wilderness Waterway
  Dept of Conservation, SHS #22, Augusta, ME 04333

Alliance for Chesapeake Bay
  410 Severn Ave, Ste 110, Annapolis, MD 21403

Alliance for Environmental Education, Inc.
  PO Box 368, The Plains, VA 22171, (703) 253-5812

American Alliance for Health
  PE & Recreation & Dance, 1900 Association Dr, Reston, VA 22091, (703) 476-3400

American Association for Botanical Gardens & Arboretums, Inc.
  PO Box 206, Swarthmore, PA 19081, (215) 328-9145

American Association for the Advancement of Science
  1333 H St NW, Washington, DC 20005, (202) 326-6400

American Association of Zookeepers, Inc.
  635 Gage Blvd, Topeka, KS 66606, (913) 272-5821

American Association of Zoological Parks & Aquariums
  Oglebay Park, Wheeling, WV 26003, (304) 242-2160

American Bass Association, Inc.
  886 Trotters Trail, Wetumpka, AL 36092, (205) 567-6035

American Birding Association
  PO Box 6599, Colorado Springs, CO 80934, (800) 634-7736

American Camping Association, Inc.
  5000 St Rd 67N, Martinsville, IN 46151, (317) 342-8456

American Cetacean Society
  PO Box 2639, San Pedro, CA 90731, (213) 548-6279

American Chemical Society
  1155 16th St NW, Washington, DC 20036, (202) 872-4600

American Council for an Energy-Efficient Economy
  1001 Connecticut Ave NW, Ste 801, Washington, DC 20036

American Environmental Management
  10960 Boatman Way, Stanton, CA 90680

American Farm Bureau Foundation
  225 Touhy Ave, Park Ridge, IL 60068, (312) 399-5700

American Farmland Trust
  1920 N St NW, Ste 400, Washington, DC 20036

American Fisheries Society
  5410 Grosvenor Lane, Bethesda, MD 20814, (301) 897-8616

Americans for the Environment
  1400 15th St NW, Washington, DC 20036, (202) 797-6665

American Forest Council
  1250 Connecticut Ave NW, Ste 320, Washington, DC 20036, (202) 463-2455

American Forests (formerly American Forestry Association)
  1516 P St NW, Washington, DC 20005, (202) 667-3300

American Geographical Society
  156 Fifth Ave, Ste 600, New York, NY 10010, (212) 242-0214

American Hiking Society
  Box 20160, Washington, DC 20041

American Horse Protection Association, Inc.
  1000 29th St NW, Ste T-100, Washington, DC 20007, (202) 965-0500

American Humane Association
  9725 East Hampden, Denver, CO 80231, (303) 695-0811

American Institute of Biological Sciences, Inc.
  730 11th St NW, Washington, DC 20001, (202) 628-1500

American Institute of Fishery Research Biologists
  15 Adamswood Rd, Asheville, NC 28803, (704) 274-7773

American Interplex Corp.
  8600 Kanis Rd, Little Rock, AR 72204

American Littoral Society, New Jersey
  Sandy Hook, Highlands, NJ 07732, (908) 291-0055

American Lung Association
  1740 Broadway, New York, NY 10019, (212) 315-8700

American Museum of Natural History
  SW Research Station, Portal, AZ 85632, (602) 558-2396

American Nature Study Society
  5881 Cold Brook Rd, Homer, NY 13077, (607) 749-3655

American Pedestrian Association
  Forest Hills Station, PO Box 624, Forest Hills, NY 11375

American Planning Association
  1776 Massachusetts Ave NW, Washington, DC 20036, (202) 872-0611

American Resources Group
  Signet Bank Bldg, Ste 210, 374 Maple Ave East, Vienna, VA 22180

American River Management Society
  Box 621911, Littleton, CO 80162, (208) 962-3245

American Rivers
  801 Pennsylvania Ave SE, Ste 400, Washington, DC 20003, (202) 547-6900

American Society of Landscape Architects
  4401 Connecticut Ave NW, Washington, DC 20008, (202) 686-ASLA

American Society of Zoologists
  104 Sirius Circle, Thousand Oaks, CA 91360, (805) 492-3585

American Trails Network
  c/o RTC Ste 400, 1325 Massachusetts Ave NW, Washington, DC 20005

American Waste Processing, Inc.
  PO Box 306, Maywood, IL 60153, (708) 681-3999

American Water Resources Association
  5410 Grosvenor Lane, Ste 220, Bethesda, MD 20814

American Wildlands Alliance
  7500 E Arapahoe Rd, Ste 355, Englewood, CO 80112, (303) 771-0380

American Youth Foundation
  Camp Merrowvista, Ossipee, NH 03864

American Youth Foundation
  PO Box 216, Three Rivers, MI 49093

American Youth Foundation
  1415 Elbridge Payne Rd, Ste 210, St. Louis, MO 63017

American Youth Foundation
  1315 Ann Ave, St. Louis, MO 63104

American Youth Hostels, District of Columbia
  733 15th St NW, Ste 840, Washington, DC 20005, (202) 783-6161

Analytical Technologies, Inc.
  5550 Morehouse Dr, San Diego, CA 92121, (619) 458-9141

Animal Protection Institute of America
  PO Box 22525, 6130 Freeport Blvd, Sacramento, CA 95822, (916) 422-1921

Animal Welfare Institute
  PO Box 3650, Washington, DC 20007, (202) 337-2333

Anniston Museum of Natural History
  PO Box 1587, Anniston, AL 36202

Antioch New England
  Graduate Environmental Studies Dept, Roxbury St, Keene, NH 03431

Appalachian Mountain Club
  Box 298, Gorham, NH 03581, (603) 466-2721

Appalachian Mountain Club
  5 Joy St, Boston, MA 02108

Appalachian Trail Conference
  PO Box 807, Harpers Ferry, WV, (304) 535-6331

Appropriate Technology Transfer for Rural Areas
  Box 3657, Fayetteville, AR 72702, (501) 442-9842

ARCC, Inc.
  PO Box 2337, Mill Valley, CA 94942

Archaeological Conservancy, The
  415 Orchard Dr, Santa Fe, NM 87501, (505) 982-3278

Arctic Institute of North America
  University Library Tower, 2500 University Dr NW, Calgary, Alta., Canada, (403) 220-7515

Arctic to Amazonia Alliance
  Box 73, Strafford, VT 05072

Arizona Department of Environmental Quality
  2005 N Central Ave, Phoenix, AZ 85004, (602) 257-2231

Arizona Department of Water Resources
  15 South 15th Ave, Phoenix, AZ 85007

Arizona Nature Conservancy, The
  Muleshoe Ranch Preserve, RR 1, Box 1542, Willcox, AZ 85643

Arizona State Parks Board
  800 W Washington, Ste 415, Phoenix, AZ 85007

Arizona-Sonora Desert Museum
  2021 North Kinney Rd, Tucson, AZ 85743

Arkansas Game & Fish Commission
  2 Natural Resource Dr, Little Rock, AR 72205, (501) 223-6300

Arkansas Nature Conservancy
  300 Spring Blvd, Rm 717, Little Rock, AR 72201

Arlington Echo Outdoor Education Center
  975 Indian Landing Rd, Millersville, MD 21108

Arlington Outdoor Education
  Science Center, House B, 869 Massachusetts Ave/AHS, Arlington, MA 02174, (617) 646-1000

Armand Bayou Nature Center
  PO Box 58828, Houston, TX 77258

Arrowhead Ranch
  Box 123, Lake Arrowhead, CA 92352, (909) 867-7041

ASCENT
  PO Box 1768, Sandpoint, ID 83864, (208) 265-2220

Aspen Center for Environment Studies
  PO Box 8777, Aspen, CO 81612

Association for Conservation Information
  PO Box 98000, Baton Rouge, LA 70898

Association of American Geographers
  1710 16th St NW, Washington, DC 20009

Association of Experiential Education
  c/o Aurora University, Aurora, IL 60506

Association of Forest Service Employees for Environmental Ethics
  Box 11615, Eugene, OR 97440

Association of New Jersey Recyclers
66 Leighton Ave, Red Bank, NJ 07701

Association of Vermont Recyclers
PO Box 1244, Montpelier, VT 05601, (802) 229-1833

ASTECH, Inc.
317 West Milton Ave, Rahway, NJ 07065, (201) 396-4455

Astrocamp
1420 Claremont Blvd, Ste 104A, Claremont, CA 91711

ATEC Associates, Inc.
PO Box 501970, Indianapolis, IN 46250, (317) 577-1761

ATC Environmental, Inc.
104 East 25th St, New York, NY 10010, (212) 353-8280

Atlantic Center for the Environment
39 South Main St, Ipswich, MA 01938, (508) 356-0038

Atmospheric & Environmental Research
840 Memorial Dr, Cambridge, MA 02139, (617) 547-6207

Atlantic Salmon Federation
PO Box 429, Saint Andrews, N.B., Canada, (506) 529-8889

Auburn University
School of Forestry, Auburn University, AL 36849

Audubon Expedition Institute
RR 1, Box 7, Mt. Vernon, ME 04352

Audubon Naturalist Society of the Central Atlantic States, Inc.
8940 Jones Mill Rd, Chevy Chase, MD 20815, (301) 652-9188

Audubon Park & Zoological Gardens
PO Box 4327, New Orleans, LA 70178

Audubon Society, National
See National Audubon Society

Aullwood Audubon Center & Farm
1000 Aullwood Rd, Dayton, OH 45414

Austin Nature Center
401 Deep Eddy, Austin, TX 78703

AWD Technologies, Inc.
15204 Omega Dr, Rockville, MD 20850, (301) 948-0040

BARC-West
USDA Agricultural Research Station, Bldg 003, Rm 403, Beltsville, MD 20705

Bard College Field Station
Hudsonia, LTD, Annadale, NY 12504

Barr Engineering Company
8300 Norman Center Dr, Minneapolis, MN 55437, (612) 832-2600

Barrie Island Environmental Education Program, South Carolina
2810 Seabrook Island Rd, John's Island, SC 29455, (803) 768-0429

Bass Anglers for Clean Water, Inc.
PO Box 17900, Montgomery, AL 36141, (205) 272-9530

Bass Research Foundation
1001 Market St, Chattanooga, TN 37402

Baxter State Park
64 Balsam Dr, Millinocket, ME 04462

Baylor School
2604 Steward Rd, Signal Mountain, TN 37377, (615) 265-4068

BC Analytical
801 Western Ave, Glendale, CA 91201, (818) 247-5737

BC Analytical
1255 Powell St, Emeryville, CA 94608, (415) 428-2300

BC Analytical
1200 Pacifico Ave, Anaheim, CA 92805

BCI
PO Box 5467, Lakeland, FL 33807, (813) 646-8591

Bear Creek Education Center
Box 880, Russellville, AL 35653

Bear Lake Wildlife Refuge
HC 69, Box 1700, Hamer, ID 83254

Bear Mountain Outdoor School
Rte 250, Hightown, VA 24444

Bear Pole Ranch Camp
32305 RCR 38, Steamboat Springs, CO 80487

Beaver Lake Nature Center
8477 East Mud Lake Rd, Baldwinsville, NY 13027

Belvedere Castle
Central Park, 830 Fifth Ave, New York, NY 10021

Bennettt & Williams, Inc.
2700 East Dublin-Granville Rd, Columbus, OH 43231

Berie, Kass & Case, NY
45 Rockefeller Plaza, New York, NY 10111

Berkshire Garden Center
Box 826, Stockbridge, MA 01267

Bermuda Biological Station
Ferry Reach GEOI, 17 Biological Station Lane, Bermuda

Berry Botanic Gardens
11505 SW Summerville Ave, Portland, OR 97219

Betsy-Jeff Penn 4-H Educational Center
Rte 13, Box 249X, Reidsville, NC 27320, (919) 249-9445

Betz Laboratories, Inc.
1 Quality Way, Trevose, PA 19053

BHA Group, Inc.
880 East 63rd St, Kansas City, MO 64133, (816) 356-8400

Big Bend Natural History Association
PO Box 68, Big Bend National Park, TX 79834

Big Five Expeditions Ltd.
2151 East Dublin–Granville Rd, Ste 215, Columbus, OH 43229

Biological Research Station
PO Box 188, Rensselaerville, NY 12147

Bio/West, Inc.
1063 West 1400 North, Logan, UT 84321, (801) 752-4202

Biosis
2100 Arch St, Philadelphia, PA 19103

Biospherics, Inc.
12051 Indian Creek Court, Beltsville, MD 20705

Black & Veatch
    8400 Ward Parkway, Kansas
    City, MO 64114
Blackfeet Community College
    Bio Science Institute, PO Box
    189, Browning, MT 59417
Block Island Bioreserve
    Box 1287, Block Island, RI
    02807
Blue Hills Reservation
    Forest Engineering, 695
    Hillside St, Milton, MA 02186
Blue Ox Forestry Service
    Rte 2, Box 194, Dodgeville, WI
    53533
Bluff Mountain Nature Preserve
    PO Box 805, Chapel Hill, NC
    27514
BMW & Associates
    9896 Bissonet, Ste 610,
    Houston, TX 77036
Boise Complex State Parks
    Lucky Peak, Veterans & Eagle
    Island, Statehouse Mail, Boise,
    ID 83720
Bok Towers Gardens
    PO Box 810, Lake Wales, FL
    33859
*Bolton Institute for a Sustainable
Future*
    4 Linden Square, Wellesley,
    MA 02181
Boojum Wilderness Institute
    PO Box 2236, Leucadia, CA
    92024
*Boston Department of Parks &
Recreation*
    1010 Massachusetts Ave, 3rd fl,
    Boston, MA 02118
Boston Harbor Island State Park
    349 Lincoln St, Hingham, MA
    02043, (617) 740-1605
Boston University Sargent Camp
    RFD 3, Windy Row,
    Peterborough, NH 03458
*Boston Urban Gardeners at the
Community Farm, Massachusetts*
    46 Chestnut Ave, Jamaica
    Plain, MA 02130, (617) 522-
    1259
Bowman's Hill Wildflower Preserve
    Box 103, Washington Crossing,
    PA 18977

Bradford Woods
    5040 State Rd 67 North,
    Martinsville, IN 46151, (317)
    342-2915
Brandywine Creek Nature Center
    PO Box 3782, Greenville, DE
    19807
Breakheart Reservation
    177 Forest St, Saugus, MA
    01906
*Breckenridge Outdoor Education
Center*
    PO Box 697, Breckenridge, CO
    80424
Briggs Associates Inc.
    400 Hingham St, Rockland,
    MA 02370, (617) 871-6040
Brookdale Community College
    The Ocean Institute, Box 533,
    Sandy Hook, NJ 07732
Bruce Company
    1100 6th St SW, Ste 515,
    Washington, DC 20024, fax
    (202) 479-1009
Bruneau Dunes State Park
    Star Rte B, Box 41, Mountain
    Home, ID 83647
Bryon Forest Preserve
    7993 N River Rd, Bryon, IL
    61010, (815) 234-8535
Bureau of Indian Affairs
    Interior South Bldg, 1951
    Constitution Ave NW,
    Washington, DC 20245.
Bureau of Indian Affairs
    Box 26567, Albuquerque, NM
    87125, (505) 766-2294
Bureau of Indian Affairs
    115 4th Ave SE, Federal Bldg,
    Aberdeen, SD 57401, (605)
    226-7550
Bureau of Indian Affairs
    5 Corporate Plaza, 3625 NW
    56th St, Oklahoma City, OK
    73112, (405) 945-6935
Bureau of Indian Affairs
    Juneau Area Office, PO Box
    25520, Juneau, AK 99802,
    (907) 586-7010
Bureau of Indian Affairs
    Phoenix Area Office, Box 10,
    Phoenix, AZ 85001, (602) 379-
    4262

Bureau of Indian Affairs
    316 N 26th St, Billings, MT
    59101, (406) 657-6381
*Bureau of Indian Affairs (DOI),
Florida*
    Star Route, Box 37, Clewiston,
    FL 33440, (813) 983-7029
Bureau of Land Management
    222 W 7th Ave, Anchorage, AK
    99513, (907) 271-5076
Bureau of Land Management
    Arizona State Office, Box
    16563, Phoenix, AZ 85011,
    (602) 640-5501
Bureau of Land Management
    Federal Office Bldg, Rm E-
    2841, 2800 Cottage Way,
    Sacramento, CA 95825, (916)
    978-4743
Bureau of Land Management
    Boise Inter-Agency Fire Center,
    3905 Vista Ave, Boise, ID
    83705, (208) 389-2446
Bureau of Land Management
    Denver Federal Center, Bldg
    50, Denver, CO 80225, (303)
    236-6452
Bureau of Land Management
    222 N 32nd St, Billings, MT,
    (406) 255-2904
Bureau of Land Management
    Box 2965, 1300 NE 44th Ave,
    Portland, OR 97208, (503)
    280-7235
Bureau of Land Management
    850 Harvard Way, Reno, NV
    89520, (702) 785-6590
Bureau of Land Management
    120 South Federal Place, Santa
    Fe, NM 87504, (505) 988-
    6030
Bureau of Land Management
    324 South State St, Ste 301,
    Salt Lake City, UT 84111, (801)
    539-4010
Bureau of Land Management
    2515 Warren Ave, Cheyenne,
    WY 82003, (307) 775-6001
Bureau of Land Management
    Eastern States Office, 350 S
    Pickett St, Alexandria, VA
    22304, (703) 22304

Bureau of Land Management
Escalante Resource Area,
Escalante, UT 84726, (801)
826-4291

Bureau of Land Management
Cedar City District Office, 176
East D. L. Sargent Dr, Cedar
City, UT 84720, (801) 586-
2401

Bureau of Land Management
Moab District Office, 82 East
Dogwood, Moab, UT 84532,
(801) 259-8193

Bureau of Land Management
PO Box 25047, Bldg 50,
Denver Federal Center, Denver,
CO 80225

Bureau of Land Management
Richfield District Office, 15
East 500 North, House Range
Resource Area, Fillmore, UT
84631, (801) 896-8221

Bureau of Land Management
Salt Lake District Office, 2370
South 2300 West, Salt Lake
City, UT 84119, (801) 977-
4300

Bureau of Land Management
20 Hamilton Court, Hollister,
CA 95023

Bureau of Land Management
Vernal District Office, 170
South 500 East, Vernal, UT
84078, (801) 539-4025

*Bureau of Land Management,
District of Columbia*
18th & C St NW, Mailstop
3619, Washington, DC 20040

Bureau of Land Management, Idaho
3380 Americana Terrace, Boise,
ID 83706, (298) 384-3245

Cable Natural History Museum
PO Box 416, Cable, WI 54821

Calgon Corporation
Pittsburgh, PA 15230, (412)
777-8000

California Academy of Sciences
Golden Gate Park, San
Francisco, CA 94118, (415)
221-5100

California Air Resources Board
PO Box 2815, Sacramento, CA
95812, (916) 445-4383

California Coastal Commission
631 Howard St, 4th fl, San
Francisco, CA 94105

California Conservation Corps
1530 Capitol Ave, Sacramento,
CA 95814, (916) 445-0307

*California Department of Air
Resources*
1234 U St, Sacramento, CA
95818

*California Department of Fish &
Game*
407 W Line St, Bishop, CA
93514

*California Department of Fish &
Game*
Endangered Plant Program,
1416 9th St, 12th fl,
Sacramento, CA 95814

*California Department of Fish &
Game*
1234 6th St, Long Beach, CA
90802

*California Department of Forestry &
Fire Protection*
1416 9th St, Sacramento, CA
94244

*California Department of Parks &
Recreation*
PO Box 942896, 1416 9th St,
Sacramento, CA 94296

*California Department of Water
Resources*
1416 9th St, Rm 303-38,
Sacramento, CA 95814

California Energy Commission
1516 9th St, MS-28,
Sacramento, CA 95814

*California Environmental Protection
Agency*
555 Capitol Mall, Ste 235,
Sacramento, CA 95814

*California Environmental Protection
Agency*
Box 2815, Sacramento, CA
95812, (916) 324-8124

California Land Management
Lake Tahoe Office, Box 14000,
South Lake Tahoe, CA 96151,
(916) 544-5994

*California League of Conservation
Voters*
965 Mission St, Ste 705, San
Francisco, CA 94103

California Marine Mammal Center
Golden Gate National
Recreation Area, Marin
Headlands, Ft. Cronkite, CA
94965

California Native Plant Society
CNPS Botanist, 909 12th St,
Ste 116, Sacramento, CA 95814

California Resources Agency, The
1416 9th St, Rm 1311,
Sacramento, CA 95814, (916)
445-5656

California Rural Water Association
1121 L St, Sacramento, CA
95814

California State Land Commission
1807 13th St, Sacramento, CA
95814, (916) 322-7801

California State University
Dept of Biology Science,
California State University,
Chico, CA 95929

California Wildlife Defenders
PO Box 2025, Los Angeles, CA
90078

Callaway Gardens
PO Box 2000, Pine Mountain,
GA 31822

*Cal-Wood Environmental Education
Resource Center*
Box 347, Jamestown, CO
80455, (303) 449-0603

*Camp Greenville Environmental
Education Center, North Carolina*
Box 370, Cedar Mountain, NC
28718, (803) 836-3291

Camp Tapawingo
Box 1353, Scarborough, ME
04070, (207) 883-7052

Campfire Club of America
230 Camp Fire Rd,
Chappaqua, NY 10514, (914)
941-0199

Canonie Environmental
800 Canonie Dr, Porter, IN
46304

Canyonlands Field Institute
Box 68, Moab, UT 84532,
(801) 259-7750

Cape Cod Academy
Box 469, Osterville, MA 02655
Cape May County Planning Board
Mechanic St, Cape May Court
House, Cape May, NJ 08210
Cape Outdoor Discovery
47 Old County Rd, East
Sandwich, MA 02537
CARE, International Employment
660 First Ave, New York, NY
10016, (212) 532-6162
Caribbean Conservation Association
Savannah Lodge, The Garrison,
St. Michael, Barbados, (809)
426-5373
*Caribbean Conservation
Corporation, Costa Rica*
Box 2866, Gainesville, FL
32602, (904) 373-6441
Carolina Raptor Center
PO Box 16443, Charlotte, NC
28297
Carrizo Plain Preserve
PO Box 3098, California
Valley, CA 93453
Carrying Capacity Network
1325 G St NW, Ste 1003,
Washington, DC 20037, (202)
879-3044
Cascade School
PO Box 9, Whitmore, CA
96096
*Castle Rock Center for
Environmental Adventure*
412 Rd 6 NS, Cody, WY 82414
Catalina Island Marine Institute
Box 795, Avalon, CA 90704,
(310) 510-1622
*Catalina Island Marine Institute,
Guided Discoveries*
Box 1360, Claremont, CA
91711
*Catskill Center for Conservation and
Development*
Dept W, Erpf House, Arkville,
NY 12406
*CDM (Camp, Dresser & McKee,
Inc.)*
1 Cambridge Center,
Cambridge, MA 02142, (617)
621-8181

CEIP Fund, Inc.
The Environmental Careers
Organization, 68 Harrison
Ave, 5th fl, Boston, MA 02111
CEIP Northwest
731 Securities Blvd, Seattle, WA
98101
Center for Marine Conservation
1725 DeSales St, Washington,
DC 20036
*Century West Engineering
Corporation*
825 Northeast Multinomah,
Portland OR 97232
CH2M Hill
PO Box 221111, Denver, CO
80222-9998
Challenge Alaska
PO Box 110065, Anchorage,
AK 99511
Charles Darwin Research Station
Isla Santa Cruz, Galapagos,
Ecuador
Charles Lindbergh Fund
Box O, Summit, NJ 07901
Charles River Watershed Association
2391 Commonwealth Ave,
Auburndale, MA 02166
Chattahoochee Nature Center
9135 Willep Rd, Rosewell, GA
30075
Chattahoochee Valley Forestry
PO Box 307, Clayton, AL
36016
Chemical Waste Management, Inc.
107 S Motor Ave, Azusa, CA
91702
Chesapeake Bay Foundation
162 Prince George St,
Annapolis, MD 21401, (301)
268-8816
Chesapeake Bay Foundation
214 State St, Harrisburg, PA
17101
*Chesapeake Bay Wildlife Sanctuary,
MD*
17308 Queen Ann Bridge Rd,
Bowie, MD 20716, (301) 390-
7010

Chesapeake Biological Lab
PO Box 38, Solomons, MD
20688
Chesapeake Research Consortium
PO Box 1120, Gloucester
Point, VA 23062
Chewonki Foundation
RR 2, Box 1200, Wiscasset, ME
04578
Chicago Botanic Garden
PO Box 400, Glencoe, IL
60022
Chippewa Nature Center
c/o Link/Erickson Associates,
148 East Main St, Midland, MI
48640
*Christordora-Manice Education
Center*
N Adams, MA 02154
Cincinnati Nature Center
Gorman Heritage Farm, 4949
Tealtown Rd, Milford, OH
45150, (513) 831-1711
Citizens for a Better Environment
88 1st St, San Francisco, CA
94105
Citizens for a Better Environment
501 2nd St, Ste 305, San
Francisco, CA 94107
Citizens for a Better Environment
407 S Dearborn, Ste 1775,
Chicago, IL 60605
Citizens for a Better Environment
1438 Vilas Ave, Madison, WI
53711
City of Boulder Venture Program
PO Box 791, Boulder, CO
80306
City of Champaign
Personnel Services Dept, 102 N
Neil St, Champaign, IL 61820
City of Chicago
Personnel Dept, 121 N LaSalle
St, Chicago, IL 60602, (312)
744-2563
City of Portland
1220 SW 5th Ave, Portland,
OR 97204

City of St. Louis
Dept of Personnel, Rm 100,
City Hall, St. Louis, MO 63103

City Park Foundation
Van Cortlandt & Pelham Bay
Parks, 1 Bronx River Parkway,
New York, NY 10462, (212)
430-1890

City Parks Foundation
Wildland Studies, 3 Mosswood
Circle, Cazadero, CA 95421,
(707) 632-5665

City Volunteer Corp., The
838 Broadway, New York, NY
10003, (212) 475-6444

*Citizen's Clearing House for
Hazardous Wastes, Virginia*
Box 6806, Falls Church, VA
22040, (703) 237-2249

*Citizen's Environmental Coalition,
New York*
33 Central Ave, Albany, NY
12210

Clean Air Engineering
500 West Wood St, Palatine, IL
60067

Clean Water Action
1320 18th St NW, Washington,
DC 20036, (202) 457-1286

Clemson University
Ecotoxicology Section,
Institute of Wildlife
Toxicology, PO Box 2278,
Clemson, SC 29632

Clemson University
Institute of Wildlife
Toxicology, 1 Tiwet Dr, PO Box
709, Clemson, SC 29670,
(803) 646-7265

CLM Services Corp.
675 Gilman St, Palo Alto, CA
94301, (415) 322-1181

Clyde Buckley Wildlife Sanctuary
1305 Germany Rd, Frankfort,
KY 40601, (606) 873-5711

CN Geotech, Inc.
PO Box 14000, Grand
Junction, CO 81502

College of the Atlantic
MDR Research Station, Allied
Whale, Bar Harbor, ME 04069

College of the Virgin Islands
Virgin Islands Coop Extension
Services, PO Box L, Kingsbell,
St. Croix, VI

College Settlement Camps, The
600 Witmer Rd, Horsham, PA
19044, (215) 542-7974

*Colorado Department of Natural
Resources*
1313 Sherman, Rm 718,
Denver CO 80203

Colorado Outward Bound School
945 Pennsylvania Ave, Denver,
CO 80203

Colorado Wildlife Federation
Box 612993, Littleton, CO
80127

*Columbia River Inter-Tribal Fisheries
Commission*
729 NE Oregon, Ste 200,
Portland, OR 97232

*Committee for Sustainable
Agriculture*
Box 1300, Colfax, CA 95713,
(916) 346-2180

Commonwealth Zoological Society
Franklin Park Zoo and Stone
Zoo, Franklin Park Zoo,
Boston, MA 02121, (617) 442-
2002

Concern, Inc.
1794 Columbia Rd NW,
Washington, DC 20009

Connecticut Audubon Society
Fairfield Nature Center, 2325
Burr St, Fairfield, CT 06430,
(203) 259-6305

Connecticut Conservation Corps
Dept of Environmental
Protection, Div of
Conservation & Preservation,
Hartford, CT 06106

*Connecticut Hazardous Waste
Management Services*
900 Asylum Ave, Ste 360,
Hartford, CT 06105

*Connecticut River Watershed
Council*
125 Combs Rd, Easthampton,
MA 01027

Conservancy, Inc., The
Naples Nature Center, 1450
Merrihue Dr, Naples, FL
33942, (813) 262-0304

Conservation Corps of Long Beach
International City Bank, PO
Box 22649, Long Beach, CA
90801

*Conservation, Environment &
Historic Preservation*
Box 18364, Washington, DC
10036

Conservation Foundation
1250 24th St NW, Washington,
DC 20037

Conservation Fund, The
1800 N Kent St, Arlington, VA
22209

Conservation International
1015 18th St NW, Ste 1000,
Washington, DC 20036

Conservation Law Foundation
62 Summer St, Boston, MA
02110

Conservation Resources, Inc.
Box 31102, Seattle, WA 98103

Consulting Environmental Engineers
100 Shield St, West Hartford,
CT 06110

Converse Environmental West
3393 Foothill Blvd, Pasadena
CA 91107

Coolidge Center for the Environment
1675 Massachusetts Ave,
Cambridge, MA 02138, (617)
864-5085

Co-op America
1850 N St NW, Ste 700,
Washington, DC 20063

*Coop Association for New England
Parks, Massachusetts*
28 Henry St, Arlington, MA
02174

*Coop Extension, University of
California, Berkeley*
2120 University Ave, 6th fl,
Berkeley, CA 94720

Cornell Laboratory of Ornithology
159 Sapsucker Woods Rd,
Ithaca, NY 18450

County of San Diego
Dept of Human Resources, 1600 Pacific Highway, Rm 207, San Diego, CA 92101

Cousteau Society, Inc., The
870 Greenbrier Circle, Ste 402, Chesapeake, VA 23320

Cox, Darrow & Owens, Inc.
1040 Kings Highway North, Cherry Hill, NJ 08034, (609) 667-3900

Coyote Creek Riparian Station
PO Box 1027, Alviso, CA 95002, (408) 262-9204

Cradlerock Outdoor Network
Box 1431, Princeton, NJ 08542

Cradlerock Outdoor Network
1313 5th St, Ste 216E, Minneapolis, MN 55414

Cready Outdoor Education Center
PO Box 5016, Brookfield, NY 06804

Creative Media/Environmental Education
9730 Manitou Place, Bainbridge Island, WA 98110

Critical Mass Energy Project
215 Pennsylvania Ave SE, Washington, DC 20003

Crow's Neck Environmental Education Center
Box 616, Tupelo, MS 38802

Crystal Fisheries, Inc.
282 Crest Dr, Soldotna, AK 99669

Crystal Run Environmental Education Center
681 Chestnut Ridge Rd, Chestnut Ridge, NY 10977

Center for American Archeology
Kampsville Archeology Center, Kampsville, IL 62053

Center for Coastal Studies
Box 1036, Provincetown, MA 02637

Center for Environmental Information
50 Whitestone Lane, Rochester, NY 14618

Center for Marine Conservation
1725 DeSales St NW, Ste 5000, Washington, DC 20036

Center for Pacific Studies
1100 Mountain Heights Way, Seattle, WA 98103

Center for Plant Conservation
125 Arborway, Jamaica Plain, MA 02130

Center for Policy Alternatives
1875 Connecticut Ave SW, Ste 710, Washington, DC 20009, (202) 387-6030

Center for Science in the Public Interest
1875 Connecticut Ave NW, Ste 300, Washington, DC 20009, (202) 332-9110

Cultural Survival
53-A Church St, Cambridge, MA 02138.

Customized Guided Excursions
PO Box 964, Hershey, PA 17033

Cuyahoga Valley Environmental Education Center
15610 Vaughn Rd, Brecksville, OH 44141, (216) 657-2054

Dade County Department of Environmental Resources
111 NW 1st St, Ste 1310, Miami, FL 33128

Dakota Resources Council
Box 1095, Dickinson, ND 59601, (701) 227-1851

Dakota Rural Action
Box 549, Brookings, SD 57006

Dames & Moore
911 Wilshire Blvd, Los Angeles, CA 90017, (213) 683-1560

Darien Nature Center
Box 1603, Darien, CT 06820, (203) 655-7459

Dartmouth Natural Resources Trust, Massachusetts
Box P-17, South Dartmouth, MA 02748

Data Contractors, Inc.
600 W 41st Ave, Ste 203, Anchorage, AK 99503

Dauphin Island Sea Lab
PO Box 369-370, Dauphin Island, AL 36528, (205) 861-2141 ext 54

Davy Crockett National Forest
Hope Center for Youth, 4115 Yeakum Blvd, Houston, TX 77006

Dawes Arboretum, The
7770 Jacksontown Rd SE, Newark, OH 43055, (614) 323-2355

De Leuw, Cather and Company
1133 15th St NW, Washington, DC 20005

Delaware Department of Natural Resources and Environmental Control
Delaware State Personnel, Townshed Bldg, Box 1401, Dover, DE 19801

Delaware Nature Society
Box 700, Hockessin, DE 19707

Defenders of Wildlife
1101 14th St NW, Ste 1400, Washington, DC 20005, (202) 682-9400

Dennison Environmental Service
600 West Cummings Park, Woburn, MA 01801

Department of Agriculture, US
Human Resources, 14th St and Independence Ave, SW, Washington, DC 20250

Department of Agriculture, Soil Conservation Service Special Examining Unit
PO Box 37636, Washington, DC 20013

Department of Energy, Environmental Affairs Division
1000 Independence Ave SW, Washington, DC 20585

Department of Energy, Oak Ridge Operations
Personnel Div, PO Box 2001, Oak Ridge, TN 37831

Desert Sun Science Center
PO Box 3399, Idyllwild, CA 92549, (909) 659-6062

Development Planning & Research Association
Div of Human Resources, PO Box 727, Manhattan, KS 66502

DFH Environmental Services, Inc.
15 North Salem St, PO Box 985, Dover, NJ 07801

Dillon Nature Center
3002 East 30th St, Hutchinson, KS 67502

Discovery Program
Rte 1, Box 42, Navasota, TX 77868, (409) 825-7175

DNR Research
Wisconsin Dept of Natural Resources, 3911 Fish Hatchery Rd, Fitchburg, WI 53711

Dodge Nature Center
1795 Charlton St West, St. Paul, MN 55118, (612) 455-4531

Dothan Landmarks Foundation
Box 6362, Dothan, AL 36302, (205) 794-3452

Dow Gardens, The
1018 West Main St, Midland, MI 48640

Duke University
School of Forestry and Environmental Studies, Durham, NC 27706

Dunn Corporation
12 Metro Park Rd, Albany, NY 12205

Dutchess Land Conservancy
Box 116, Stanford, NY 12581

Dvorak Expeditions
17921-Z, U.S. Highway 285, Nathrop, CO 81236

Dynamac Corporation
2275 Research Blvd, Rockville, MD 20850

E Magazine
Box 5098, Westport, CT 06881

Eagle Services
500 North Franklin Turnpike, Ramsey, NJ 07446

Earth Care Paper Company
Dept EO, PO Box 3335, Madison, WI 53704

Earth First!
PO Box 1031, Redway, CA 95560

Earth Island Institute
300 Broadway, Ste 28, San Francisco, CA 94133

Earth Reach Environment Consultants
Rte 1, Box 178A, Ferrum, VA 24088

Earth Technology Corporation
100 W. Broadway, Ste 5000, Long Beach, CA 90802, (310) 495-3337 ext 2588

Earthwise
2F Brookside Condos, Brattleboro, VT 05301

Earthworm, Inc.
175 South St, Boston, MA 02111

East Bay Conservation Corps
1021 3rd St, Oakland, CA 94607

East Bay Regional Park District
11500 Skyline Blvd, Oakland, CA 94619

East Bay Regional Park District
Personnel Dept, 2950 Peralta Oaks Court, Oakland, CA 94605, (510) 531-9300 exts 4600, 4604

Eastconn
376 Hartford Turnpike, North Windham, CT 06256

Ebasco Environmental
see Enserch Environmental Corp

EC Jordan, Inc.
Human Resources Dept, PO Box 7050 DTS, Portland, ME 04112

Echo Hill Outdoor School
Worton Post Office, Kent County, Worton, MD 21678, (301) 348-5303

Echohorizons
22601 South West 152nd Ave, Goulds, FL 33170

Eckerd Family Youth Alternatives
Box 7450, Clearwater, FL 34618, (800) 222-1473

Ecologic
PO Box 1514, Antigonish, N.S., Canada

Ecological Restoration & Management
303 Allegheny Ave, Towson, MD 21204

Ecology and Environment, Inc.
Buffalo Corporate Center Three, Lancaster, NY 14086

Ecology Center
1403 Addison St, Berkeley, CA 94705

Ecosummer Expeditions
1516-B Duranleau St, Vancouver, B.C., Canada

Ecosystems Research Unit
Rte 6, Box 1877, Naples, FL 33964

Ecotourism Society
PO Box 755, N Bennington, VT 05257

Edudex Associates
Kettle Moraine Division, 604 2nd Ave, West Bend, WI 53095

Edwards Camp & Conference Center
1275 Army Lake Rd, PO Box 16, East Troy, WI 53120

EG & G Energy Measurements
Santa Barbara Operations, 130 Robin Hill Rd, Goleta, CA 93117

EHPC
801 South Church St, Mt. Laurel, NJ 08054

El Paso County Parks
2002 Creek Crossing, Colorado Springs, CO 80906, (719) 520-6375

Elachee Nature Science Center
311 Green St, Station SE, Gainesville, GA 30501

EMCON Southeast
8021 Phillips Hwy, Ste 12, Jacksonville, FL 32256

Emilcott Associates, Inc.
CIH at 37 Kings Rd, Madison, NJ 07940

EnCap, Inc.
400 E Hilcrest, Ste 240, Dekalb, IL 60115

Endangered Species Coalition
666 Pennsylvania Ave SE, Washington, DC 20003

Energy Coordination Agency of Philadelphia
1501 Cherry St, Philadelphia, PA 19102

Engineering-Science, Inc.
100 Walnut St, Pasadena, CA 91124

Enserch Environmental Corp, Western Headquarters
Human Resources, 3000 W MacArthur Blvd, Santa Ana, CA 92704

Enserch Environmental Corp, Northeastern Headquarters
160 Chubb Ave, Lyndhurst, NJ 07071

Enserch Environmental Corp, Southeastern Headquarters
145 Technology Park, Norcross, GA 30092

Enviro Media
PO Box 1016, Chapel Hill, NC 27514

Enviros Principal Environmental Engineers
5808 Lake Washington Blvd, NE, Kirkland, WA 98033

Envirodyne Engineers, Inc.
Human Resources Dept, 168 N Clinton St, Chicago, IL 60661

Environ
210 Carnegie Center, Ste 201, Princeton, NJ 08540

ENSR Consulting & Engineering
Western Region, 1220 Avenida Acaso, Camarillo, CA 93012

ENSR
Central Region, 3000 Richmond Ave, Houston, TX 77098

ENSR
Eastern Region, 35 Nagog Park, Acton, MA 01720

Envirologic Data
Human Resources Dept, 295 Forest Ave, Portland, ME 04101

Environmation Group, Inc., Colorado
1110 Clubview Terrace, Ste. J81, Ft. Collins, CO 80524

Environment Laboratories, Inc.
4083 Main St, Bridgeport, CT 06606

Environment Quality Lab, Inc.
1009 Tamiami Trail, Port Charlotte, FL 33953

Environment Resource Center
1800 Business Park Dr, Clarksville, TN 37040

Environment Science & Engineering, Inc.
1099 West Grand River, Williamston, MI 48895

Environmental Science & Engineering, Inc.
300 Hamilton Bldv, Ste 330, Peoria, IL 61602

Environment Science Associates
760 Harrison St, San Francisco, CA 94107

Environment Support Center, The
1905 Queen Anne Ave N, Ste 126, Seattle, WA 98109

Environmental Action
6930 Carroll Ave, Ste 600, Takoma Park, MD 20912

Environmental & Chemical Services
Box 1393, Aiken, SC 29801

Environmental & Energy Study Institute
122 C St, Ste 700, Washington, DC 20001, (202) 628-1400

Environmental Career Center, CO
313 Rapidan Rd, Hampton, VA 23669, (804) 851-4470

Environmental Careers Organization
286 Congress St, 3rd fl, Boston, MA 02210

Environmental Concern, Inc.
Education Dept, PO Box P, St. Michaels, MD 21663, (301) 745-9620

Environmental Defense Fund
1875 Connecticut Ave NW, Ste 1016, Washington, DC 20009, (202) 387-3500

Environmental Defense Fund, New York
257 Park Ave South, New York, NY 10010, fax (212) 505-2375

Environmental Defense Fund, Texas
1800 Guadalupe, Austin, TX 78701

Environmental Education Center, The
2 Thunderbird Lane, Clover, SC 29710, (803) 831-2121

Environmental Fund for Georgia
Box 78262, Atlanta, GA 30357

Environmental Fund of Washington
1305 4th Ave, Ste 512, Seattle, WA 98101

Environmental Health Association
135 Raritan Center Parkway, Edison, NJ 08625

Environmental Law Institute
1616 P St NW, Ste 200, Washington, DC 20036

Environmental Learning Center
61 Mitaivan Lake Rd, Isabella, MN 55607

Environmental Lobbyist
PO Box 309, Albany, NY 12201

Environmental Planning Lobby
53 Central Ave, Albany, NY 12210

Environmental Protection Agency, US, Hdqrtrs
Human Resources, 401 M St SW, Washington, DC 20460

Environmental Protection Agency, Region 1
John F. Kennedy Federal Bldg, Boston, MA 02203

Environmental Protection Agency, Region 2
26 Federal Plaza, New York, NY 10278

Environmental Protection Agency, Region 3
841 Chestnut St, Philadelphia, PA 19107

Environmental Protection Agency, Region 4
345 Courtland St NE, Atlanta, GA 30365

Environmental Protection Agency, Region 5
230 S. Dearborn, Chicago, IL 60604

Environmental Protection Agency, Region 6
1445 Ross Ave, 12th fl, Ste 1200, Dallas, TX 75202

Environmental Protection Agency, Region 7
725 Minnesota Ave, Kansas City, KS 66101

Environmental Protection Agency, Region 8
999 18th St, Ste 500, Denver, CO 80202

Environmental Protection Agency, Region 9
1235 Mission St, San Francisco, CA 94103

Environmental Protection Agency, Region 10
1200 Sixth Ave, Seattle, WA 98101

Environmental Protection Agency
Chesapeake Bay Program, Box 1280, Solomons, MD 20688

Environmental Protection Agency
Human Resources Office, PO Box 98516, Las Vegas, NV 89193

Environmental Resource Center, Idaho
Box 819, Ketchum, ID 83340, (208) 726-4333

Environmental System Analysis
708 Melvin Ave, Annapolis, MD 21401

Environmental Toxicology International
600 Steward St, Ste 700, Seattle, WA 98101

Environosphere Company
160 Chubba Ave, Lyndhurst, NJ 07071

Erwin Potter & Associate, Environmental Systems Planning Ltd.
PO Box 839, Sandy Hook, CT 06482

Escalante Canyon Outfitters
PO Box 325, Boulder, UT 84716

ESE Biosciences Group
3208 Spring Forest Rd, ·Raleigh, NC 27604

Everglades National Park
PO Box 279, Homestead, FL 33030, (305) 247-6211 ext 210

Evergreen Nursery Company, Inc.
5027 County TT, Sturgeon Bay, WI 54235

Everything Earthly
12211 Paradise Village Parkway 173, Phoenix, AZ 85032

Executive Expeditions
131 Village Parkway, Ste 4, Marietta, GA 30067

Fairchild Tropical Garden
11935 Old Cutler Rd, Miami, FL 33156

Fairfax County Park Authority
Personnel Dept, 4130 Chain Bridge Rd, Fairfax, VA 22030

Fairview Gardens Farm
598 North Fairview Ave, Goleta, CA 93117

Fairview Lake Environmental Education Center
Fairview Lake YMCA, RD 5, Box 5375, Newton, NJ 07860

Farm & Wilderness
Outdoor Education Program, Plymouth, VT 05056, (802) 422-3761

Farm Animal Reform Movement
10101 Ashburton Lane, Bethesda, MD 20817, (301) 530-1737

Farm Sanctuary
Box 150, Watkins Glen, NY 14891, (607) 583-2225

Farmland Wildlife Research Group
Rte 1, Madelia, MN 56062

Farr Company
2221 Park Place, El Segundo, CA 90245

Federal Energy Regulatory Commission
Personnel Operations, 941 W Central Capital St, Rm 3400, Washington, DC 20246

Fernwood Nature Center
13988 Range Line Rd, Niles, MI 49120, (616) 695-6491

Field Museum of Natural History
Personnel Dept, Roosevelt Rd at Lake Shore Dr, Chicago, IL 60605

Field Resource Group
1 Robert Lane, Unit B, Glen Head, NY 11545, (516) 759-7891

Fish & Wildlife Service, US
Div of Personnel Management, 1849 C St NW, Mail Stop ARLSQ-100, Washington, DC 20240, (703) 358-1743

Fish & Wildlife Service, US
Div of Personnel Management, 4401 N Fairfax Dr, Rm 100, Arlington, VA 22203

Fish & Wildlife Service, US
Pacific Regional Office, 911 NE 11th Ave, Portland, OR 97232

Fish & Wildlife Service, US
Southwest Regional Office, 500 Gold Ave, SW, Rm 3018, Albuquerque, NM 87102

Fish & Wildlife Service, US
Great Lakes-Big Rivers Regional Office, 1 Federal Dr, Fort Snelling, MN 55111

Fish & Wildlife Service, US
Southeast Regional Office, Russell Federal Bldg, 75 Spring St SW, Atlanta, GA 30303

Fish & Wildlife Service, US
Northeast Regional Office, 300 Westgate Center Dr, Hadley, MA 01035

Fish & Wildlife Service, US
Mountain-Prairie Regional Office, 134 Union Blvd, PO Box 25486, Denver Federal Center, Denver, CO 80225

Fish & Wildlife Service, US
Alaska Regional Office, 1011 E Tudor Rd, Anchorage, AK 99503

Fish & Wildlife Service, US
Research and Development Centers Hdqrtrs, R&D Regional Office, 1849 C St NW, Washington, DC 20240

Five Rivers Environmental Education Center
Game Farm Rd, Delmar, NY 12054

Florida Audubon Research Headquarters
115 Indian Mound Trail, Tevernier, FL 33070

Florida Conservation Foundation, Inc.
1251 Miller Ave, Winter Park, FL 32789

Florida Department of Agricultural & Consumer Services
5745 S Florida Ave, Lakeland, FL 33813, (813) 648-3163

*Florida Department of
Environmental Protection*
  3900 Commonwealth Blvd,
  Tallahassee, FL 32399
Florida Division of Forestry
  3125 Conner Blvd, Tallahassee,
  FL 32399, (904) 488-7247
*Florida Division of Recreation &
Parks*
  3900 Commonwealth Blvd,
  Tallahassee, FL 32399
*Florida Game & Fresh Water Fish
Commission*
  620 South Meridian St,
  Tallahassee, FL 32399
Florida Marine Research Institute
  100 8th Ave SE, St. Petersburg,
  FL 33701
Florida Natural Areas Inventory
  1018 Thomasville Rd, Ste
  200C, Tallahassee, FL 32303
Florida Oceanographic Society
  890 NE Ocean Blvd, Stuart, FL
  34996
Flat Rock Brook Nature Center
  443 Van Nostrand Ave,
  Englewood, NJ 07631, (201)
  567-1265
*Flat Rock River Environmental
Education Center*
  RR 1, Box 102, St. Paul, IN
  47272
Foellinger-Freimann Botanical
  1100 South Calhoun St, Ft.
  Wayne, IN 46802
Fontenelle Forest Association
  8308 Hickory, Omaha, NB,
  68124
Foothill Horizon
  21925 Lyons Bald Mountain
  Rd, Sonora, CA 95370
*Foothills Land Conservancy,
Tennessee*
  Box 6069, Maryville, TN
  37802, (615) 584-6133
*Forest Preserve District of DuPage
County*
  Box 2339, Glen Ellyn, IL
  60138, (708) 790-4900
Foth & Van Dyke
  2737 South Ridge Rd, Green
  Bay, WI 54307

Forest Service, US
  Hdqrtrs, Human Resources,
  PO Box 96090, Washington,
  DC 20090-6090
Forest Service, US
  Region 1, Federal Bldg, PO Box
  7669, Missoula, MT 59807
Forest Service, US
  Region 2, 740 Sims St.
  Lakewood, CO 80401
Forest Service, US
  Region 3, Federal Bldg, 517
  Gold Ave SW, Albuquerque,
  NM 87102
Forest Service, US
  Region 4, 324 25th St, Ogden,
  UT 84401
Forest Service, US
  Region 5, 630 Sansome St, San
  Francisco, CA 94111
Forest Service, US
  Region 6, 333 SW 1st Ave,
  Portland, OR 97208
Forest Service, US
  Former Region 7 was absorbed
  into other regions
Forest Service, US
  Region 8, Ste 800, 1720
  Peachtree Rd NW, Atlanta, GA
  30367
Forest Service, US
  Region 9, 310 W Wisconsin
  Ave, Rm 500, Milwaukee, WI
  53203
Forest Service, US
  Region 10, Federal Office Bldg,
  Box 21628, Juneau, AK 99802
Forest Trust
  PO Box 519, Santa Fe, NM
  87504
Foundation for Field Research
  PO Box 2010, 787 South Grade
  Rd, Alpine, CA 92001, (619)
  445-9264
*Four Corners School of Outdoor
Education*
  East Rte, Monticello, UT 84535
Frank Orth & Associates
  10900 NE 4th St, Ste 930,
  Bellevue, WA 98004
Frederic R. Harris, Inc.
  300 E 42nd St, New York, NY
  10017

Freese & Nichols, Inc.
  811 Lamar St, Ft. Worth, TX
  76102
Friends of the Earth
  1025 Vermont Ave NW,
  Washington, DC 20005, (202)
  544-2600
Friends of the River
  909 12th St, #207,
  Sacramento, CA 95814
Friends of the Sea Otter
  Box 221220, Carmel, CA
  93922
4-H Environment Program
  PO Box 748, Oriskany, NY
  13424
*4-H Farley Outdoor Education
Center*
  Box 97, Forestdale, MA 02644,
  (508) 477-0181
Fugro-McClelland, Inc.
  6 Maple St, PO Box 780,
  Northboro, MA 01532
Fund for Animals, Inc.
  200 W 57th St, New York, NY
  10019.
GAI Consultants, Inc.
  570 Beatty Rd, Monroeville, PA
  15146
General Testing Corporation
  710 Exchange St, Rochester, NY
  14608
Geological Survey, US
  National Center, Reston, VA
  22092
Geo Research, Inc.
  2815 Montana Ave, Billings,
  MT, 59101
*George Miksch Sutton Avian
Research Center*
  PO Box 2007, Bartlesville, OK
  74005, (918) 336-7778
Georgia Conservancy, The
  711 Sandtown Rd, Savannah,
  GA 31410
*Georgia Department of Natural
Resources*
  205 Butler St SE, Ste 1352,
  Atlanta, GA 30334, (404) 656-
  7092

GeoResearch
PO Box 90911, Long Beach, CA 90809

Geraghty & Miller, Inc.
125 East Bethpage Rd, Plainview, NY 11803

German Marshall Fund of the United States
11 Dupont Circle NW, Ste 750, Washington, DC 20436

Gildea Resource Center
930 Miramonte Dr, Santa Barbara, CA 93109

Glacier Institute
PO Box 1457, Kalispell, MT 59903, (406) 756-3911

Glen Helen Nature Preserve
795 Livermore St, Yellow Springs, OH 45387

Glenwood Wilderness Camp
Glenwood Mental Health Services, 101 Glenwood Lane, Birmingham, AL 35242

Global Resource Consultants
9501 Lomond Dr, Manassas, VA 22110

Global Resources Information Clearinghouse
RD 1, Box 185, Valley Falls, NY 12210

Global Rivers Environmental Education Network
216 S State St, Ste 4, Ann Arbor, MI 48104

Golden Gate National Park Association
Ft. Mason, Bldg 201, San Francisco, CA 94123, (415) 331-0730

Gorilla Foundation
Box 620-530, Woodside, CA 94062

Great Lakes Commission
400 4th St, Ann Arbor, MI 48103

Great Smoky Mountain Institute
Rte 1, Box 700, Towsend, TN 37882, (615) 448-6709

Green Corps Summer Teen Action
1109 Walnut St, 4th fl, Philadelphia, PA 19107, (215) 829-1760

Green Environmental Services, Inc.
3950 River Ridge Dr N, Cedar Rapids, IA 52402

Green Isle Environmental Service
410 11th Ave S, Hopkins, MN 55343

Green Mountain Club
RR 1, Box 650, Rte 100, Waterbury, VT 05677

Green Seal
1250 23rd St NW, Ste 275, Washington, DC 20037

Greenkill Outdoor Environmental Education Center
YMCA of Greater New York, Huguenot, NY 12746, (914) 856-4382

Greenpeace
462 Broadway, 6th fl, New York, NY 10013, (212) 941-0994

Greenpeace USA
1436 U St NW, Washington, DC 20009

Greenpower Farm
PO Box 624, Weston, MA 02193

Gresham, Smith & Partners
3310 West End Ave, Nashville, TN 37203

Griffith Center, The
1546 Cole Blvd, Ste 225, Golden, CO 80401

Groundwater Technology, Inc.
100 River Ridge Dr, Norwood, MA 02062

Gunflint Wilderness Camp
Steamboat Lake, LaPorte, MN 56461

Gurr & Associates, Inc.
2000 E Edgewood Dr, Ste 102, Lakeland, FL 33803

Habitat Institute for the Environment
10 Juniper Rd, Box 136, Belmont, MA 02178

Hagley Museum & Library
Human Resources Dept, PO Box 3639, Wilmington, DE 29807

Halliburton NUS Environmental Corp
EA17ss, Savannah River Center, 900 Trail Ridge Rd, Aiken, SC 29803

Hamilton County Park District
10245 Winton Rd, Cincinnati, OH 45231

Hanson Engineers, Inc.
1525 S 6th St, Springfield, IL 62703

Harding Lawson Associates
Corporate Staffing, Dept Earth Work, PO Box 578, Novato, CA 94948

Hargis+Associates, Inc.
2223 Avenida De La Playa, Ste 300, La Jolla, CA 92037

Harvard University
OER, 26 Oxford St, Cambridge, MA 02138

Hatcher & Eiland, Inc.
3901 Roswell Rd NE, Ste 200, Marietta, GA 30062

Hawaii Department of Land & Natural Resources
Box 621, Honolulu, HI 96809

Hawk Mountain Sanctuary Association
Rte 2, Kempton, PA 19529

Hawthorne Valley
RD 2, Box 225, Ghent, NY 12075

Hayward Area Recreation & Park District
1099 East St, Hayward, CA 94541

Hazmat Environmental Group, Inc.
New Village Industrial Park, Buffalo, NY 14218

Hazmat Training
Columbia Business Center 6, Columbia, MD 21045

Headlands Institute, The
Yosemite National Institute, Golden Gate National Recreation Area, Sausalito, CA 94965

Heart of America
4556 University Way, Ste 201, Seattle, WA 98103

Hidden Valley Camp
Freedom, ME 04941, (207) 342-5177

High Point Environmental Education
1228 Penny Rd, High Point, NC 27260

Himalaya Trek & Wilderness Exchange
1900 8th St, Berkeley, CA 94710

Hillman Environmental Company
1089 Cedar Ave, Union, NJ 07083

Hitchcock Center for the Environment
525 S Pleasant St, Amherst, MA 01002

Hixon Forest Nature Center
2702 Quarry Rd, La Crosse, WI 54601

Holden Arboretum
9500 Sperry Rd, Mentor, OH 44060, (216) 256-1110

Homeward Bound
Box F, Brewster, MA 02631

Honey Creek Environmental Education Center
Rte 1, Box 94, Waverly, GA 31565

Honey Hollow Environmental Education Center
RD 1, Box 263A, New Hope, PA 19006

Hopi Tribe
Personnel Dept, PO Box 123, Kykotsmovi, AZ 86039

Horizons for Youth
121 Lakeview St, Sharon, MA 02067

Horn Point Environmental Labs
PO Box 775, Cambridge, MD 21613

Horticulture Society of New York
128 West 58th St, New York, NY 10019

Houston Arboretum & Nature Center
4501 Woodway Dr, Houston, TX 77024

HOWL Wildlife Center
Box 1037, Lynwood, WA 98046, (206) 743-3845

Hoyt Farm Park Preserve
PO Box 220, Commuck, NY 11725

Hudson Guild
441 W 25th St, Rm 108, New York, NY 10001

Hudson River Sloop Clearwater
112 Market St, Poughkeepsie, NY 12601

Humane Society of the United States
2100 L St NW, Washington, DC 20037, (202) 452-1100

Humboldt State University
College of Natural Resources, Arcata, CA 95521, (707) 826-3561

Huntsman Marine Science Center
Brandy Cove Rd, St. Andrews, N.B., Canada

Hurricane Island Outward Bound
Rte 2, Box 237E, Yulee, FL 32097, (904) 261-4021

Idaho Department of Fish & Game
600 S Walnut, PO Box 25, Boise, ID 83707

Idaho Department of Lands
1215 W State St, Boise, ID 83720, (208) 334-0200

Idaho Department of Parks & Recreation
Statehouse Mail, Boise, ID 83720, (208) 327-7444

Idaho Department of Water
1301 West Orchard, Boise, ID 83720

Ijams Audubon Nature Center
2915 Island Home Ave, Knoxville, TN 37920

Illinois Department of Conservation
524 S 2nd St, Springfield, IL 62701

Illinois Department of Energy & Natural Resources
325 W Adams St, Rm 300, Springfield, IL 62704

Illinois Environmental Protection Agency
2200 Churchill Rd, Springfield, IL 62794

Illinois Natural History Survey
607 East Peabody Dr, Champaign, IL 61820

Indian Rock Nature Preserve
Wolcott Rd, Bristol, CT 06010

Indiana Department of Environmental Management
105 South Meridian St, Chesapeake Bldg, Indianapolis, IN 46225

Indiana Department of Natural Resources
Human Resources Div, 402 Washington, Rm 261, Indianapolis, IN 46204

Innerquest
220 Queen St NE, Leesburg, VA 22075

Institute for Earth Education
Box 288, Warrenville, IL 60555

Institute for Earth Education
Cedar Cove, Greenville, WV 24945

Institute for Global Communication
3228 Sacramento St, San Francisco, CA 94115

Institute for Local Self-reliance
2425 18th St NW, Washington, DC 20009, (202) 232-4108

Institute of Ecosystem Studies
New York Botanical Garden, Box AB, Millbrook, NY 12545

Integrated Power Corporation
7524 Standish Place, Rockville, MD 20855

International Crane Foundation
E-11376 Shady Lane Rd, Baraboo, WI 53913, (608) 356-9462

International Fund for Animal Welfare
PO Box 193, 169 Main St, Yarmouth Port, MA 02675, (508) 362-4944

International Journal of Air Pollution Control & Hazardous Waste Management
Air Pollution Control Association, PO Box 2861, Pittsburgh, PA 15230, (412) 232-3444

International Mountain Biking Association
Box 41203, Los Angeles, CA 90041

International Osprey Foundation
PO Box 250, Sanibel Island, FL 33957

International Rivers Network
Tides Foundation, 301 Broadway, Ste B, San Francisco, CA 94133

International Society of Tropical Foresters
5400 Grosvenor Lane, Bethesda, MD 20814

International Technology Corporation
2355 Main St, Ste 100, Irvine, CA 92714.

International Wildlands Research
109 Howard St, Petaluma, CA 94952

International Wildlife Coalition
634 N Falmouth, PO Box 388, N Falmouth, MA 02556

Iowa Department of Natural Resources
Wallace State Office Bldg, Des Moines, IA 50319

Irvine Consulting Group
15 Mason, Irvine, CA 92718

Isaac W. Bernheim Foundation
Bernheim Forest Arboretum, Clermont, KY 40110

Isaacson, Miller, Gilvar & Bou
105 Chauncy St, Boston, MA 02111

Izaak Walton League of America
1401 Wilson Blvd, Level B, Arlington, VA 22209

Jackson Community College
Dahlem Environmental Education Center, 7117 S Jackson Rd, Jackson, MI 49201

Jackson Hole Land Trust
PO Box 2897, Jackson, WY 83001

Jacobs Engineering Group
251 S Lake Ave, Pasadena, CA 91101

Japan Environmental Exchange
PO Box 3322, Santa Cruz, CA 95063

Jason Cortell & Associates
244 Second Ave, Waltham, MA 02154

Jekyll Island 4-H Center
201 S Beachview Dr, Jekyll Island, GA 31527, (912) 635-4117

John Dorr Nature Lab
Box 290, Washington Depot, CT 06794

John James Audubon State Park
Box 576, Henderson, KY 42420

John Pennekamp Coral Reef Park
PO Box 487, Key Largo, FL 33037

Johnson Control World Services
Dept 137, 401 Wynn Dr, Huntsville, AL 35805, (800) 950-6990 ext 194

Jones & Jones
105 South Main St, Seattle, WA 98104, (206) 624-5702

Jordan Associates
90 New Montgomery St, Rm 410, San Francisco, CA 94105

Joy Outdoor Education Center
PO Box 157, Clarksville, OH 45113

Jug Bay Wetlands Sanctuary
1361 Wrighton Rd, Lothian, MD 20711, (410) 741-9330

Kalamazoo Nature Center
7000 N Westnedge Ave, Kalamazoo, MI 49007, (616) 381-1574

Kansas Department of Wildlife & Parks
Personnel Office, Rte 2, Box 54A, Pratt, KS, 67124

Kansas Natural Resources Council
1516 Topeka Blvd, Topeka, KS 66612

Keewaydin Environmental Education Center
Box 77, Ripton, VT 05766, (802) 388-4082

Kent Mountain Adventure Center
PO Box 835, Keystone, CO 80517

Kemron Environmental Services
7926 Jones Branch Dr, McLean, VA 22102

Kentucky Department of Fish and Wildlife Resources
#1 Game Farm Rd, Frankfort, KY 40601

Kentucky Department of Parks
Div of Recreation, Capital Plaza Tower, 11th fl, 500 Mero St, Frankfort, KY 40601, (502) 564-2172

Kentucky State Nature Preserve
407 Broadway, Frankfort, KY 40601

Kern Environmental Program
5801 Sundale Ave, Bakersfield, CA 93309

Kern Outdoor Education Center
5291 SR 350, Oregonia, OH 45054, (513) 932-3756

Kerr Center for Sustainable Agriculture
Box 588, Poteau, OK 74953, (918) 647-9123

Keystone Science School
Box 70, Montezuma Rte, Dillon, CO 80435, (303) 468-5824

Kimball Camp YMCA Nature Center
4444 Long Lake Rd, Reading, MI 49274

King's Mark Resource Conservation & Development Area
322 N Main St, Wallingford, CT 06492, (203) 284-3663

Kleinfelder
2121 N California Blvd, Rm 570, Walnut Creek, CA 94596

Laboratory Resources
363 Old Hook Rd, Westwood, NJ 07675

Lake Forest Preserve
Ryerson Conservation Areas, 21950 Riverwoods Rd, Deerfield, IL 60015, (708) 948-7750

Lake Pontchartrain Basin Foundation, Louisiana
Box 6965, Metairie, LA 70009

Lancaster Laboratories
2425 New Holland Pike, Lancaster, PA 17601

Land Between the Lakes
Professional Development, 100 Van Morgan Dr, Golden Pond, KY 42211, (502) 924-5602

Land Institute, The
2440 E Walter Well Rd, Salina, KS 67401, (913) 823-5376

Land Trust Alliance
900 17th St NW, Ste 410, Washington, DC 20006

Land Trust Exchange
1017 Duke St, Alexandria, VA 22314

Lanphere-Christensen Dunes Preserve
6800 Lanphere Rd, Arcata, CA 95521

LaPorte County Parks, Indiana
LaPorte County Park Board Office, 5th level, County Complex, LaPorte, IN 46350, (219) 326-6808 ext 271

Law Engineering & Environmental Services
114 Town Park Dr, Kennesaw, GA 30144

Lawrence Hall of Science
Summer Camp of California, Berkeley, CA 94720

Lawrence Livermore National Laboratory
PO Box 808, L410, Livermore, CA 94550

Leadership Education & Development
22 Spring St, Box 275, Williamstown, MA 01267

League of Conservation Voters
1707 L St NW, Ste 550, Washington, DC 20036

League of Women Voters Education Fund
Human Resources, 1730 M St NW, Washington, DC 20036

Legacy International
Rte 4, Box 265-A, Bedford, VA 24523, (703) 297-5982

Lehigh Valley Conservancy
601 Orchid Place, Emmaus, PA 18049

Lenox Hill Camp
Box 400, Bantam, CT 06750

Lighthawk
Box 8163, Santa Fe, NM 87504, (505) 982-9656

Lincoln Conservation Commission
PO Box 353, Lincoln Center, MA 01773

Lincoln Filene Center at Tufts University
169 Holland St, Somerville, MA 02144

Lindsley Outdoor Center
2425 Rte 168, Georgetown, PA 15043, (412) 899-2100

Linwood-MacDonald Environmental Education Center
RD 2, Box 268, Branchville, NJ 07826

Lion Meadows Wilderness Camp
27211 Henry Mayo Dr, Valencia, CA 91355, (805) 257-2490

Little Pond Camps
Township 7, Range 10, PO Box 1269, Greenville, ME 04441

Little Traverse Conservancy
3264 Powell Rd, Harbor Springs, MI 49740

Living Tree Center
Box 10082, Berkeley, CA 94709, (510) 420-1440

Lloyd Center for Environmental Studies
PO Box 7037, S Dartmouth, MA 02748, (508) 990-0505

Long Branch Environmental Education Center
Rte 2, Box 132, Leicester, NC 28748

Long Lake Conservation Center
Rte 2, Box 2550, Palisade, MN 56469, (218) 768-4653

Longacre Expeditions
RD 3, Box 106, Newport, PA 17074, (717) 567-6790

Los Angeles County Department of Arboreta and Botanical Gardens
301 N Baldwin Ave, Arcadia, CA 91007, (818) 821-3234

Los Angeles County Outdoor Science School
9300 E Imperial Hwy, Downey, CA 90242, (310) 922-6334

Los Angeles Unified School District
Office of Outdoor Education, Student Auxiliary Services Branch, 5607 Capistrano Ave, Woodland Hills, CA 91367

Lost Valley Education Center
88868 Lost Valley Lane, Dexter, OR 97431, (503) 937-3351

Louisiana Department of Wildlife and Fisheries
PO Box 98000, Baton Rouge, LA 70898

Louisiana Office of State Parks, Department of Culture, Recreation & Tourism
PO Box 44426, Baton Rouge, LA 70804

Louisiana Department of Environmental Quality
PO Box 82231, Baton Rouge, LA 70884

Lower Rio Grande Valley Nature Center
Box 8125, Weslaco, TX 78596

Lummi Indian Business Council
2105 Highway 20, Bellingham, WA 98226

Maine Appalachian Trail Club
15 Westwood Rd, Bangor, ME 04401

Maine Audubon Society
118 US Rte 1, Box 6009, Falmouth, ME 04105

Maine Bureau of Public Lands
State House, Station 4, Augusta, ME 04333

Maine Coast Heritage Trust
PO Box 426, Northeast Harbor, ME 04662

Maine Conservation School
RR2, Box 340, Bryant Pond, ME 04219

Maine Department of Conservation
State House Station 22, Augusta, ME 04333

Maine Department of Environmental Protection
State House Station 17, Augusta, ME 04333

Maine National High Adventure
PO Box 607, Howland, ME 04448, (207) 732-4845

Maine People Alliance
359 Main St, Bangor, ME 04401

Management Recruiters of Richmond
PO Box 263, Richmond, KY 40475, (606) 624-3535

Manice Education Center
666 Broadway, 9th fl, New York, NY 10012, (212) 353-2052

Manitoga, Inc.
PO Box 350, Garrison, NY 10524, (914) 424-3812

Mantech International Corporation
12015 Lee Jackson Hwy S, Fairfax, VA 22033

Maple Rock Farm
Nature's Classroom, RFD 1, Box 410, Southbridge, MA 01550

Marine Conservation Corps, California
Box 150089, San Rafael, CA 94915

Marine Science Consortium
Box 16, Enterprise St, Wallops Islands, VA 23337, (804) 824-5636

Marine Sciences Under Sails
PO Box 3994, Hollywood, FL 33083

Maritime Center, The
10 North Water St, Norwalk, CT 06854, (203) 852-0700

Martha's Vineyard Land Bank Commission
Box 2057, Edgartown, MA 02539, (508) 627-7141

Maryland Department of Natural Resources
Tawes Office Bldg, Annapolis, MD 21401

Maryland Department of the Environment
2500 Broening Hwy, Baltimore, MD 21224

Maryland Environmental Trust
275 West St, Ste 322, Annapolis, MD 21401

Maryland-National Capital Park & Planning Commission
6609 Riggs Rd, Hyattsville, MD 20782

Mashomack Preserve
PO Box 410, Shelter Island, NY 11964

Massachusetts Audubon Society, Inc.
208 South Great Rd, Lincoln, MA 01773, (617) 259-9500 ext 7804

Massachusetts Audubon Society
3 Joy St, Boston, MA 02108

Massachusetts Bay Marine Studies Consortium
PO Box 660, Boston, MA 02125

Massachusetts Coastal Resources, C.R.A.B
100 Cambridge St, Boston, MA 02108

Massachusetts Department of Environmental Management
Div of Forest & Parks, 100 Cambridge St, Boston, MA 02202, (617) 272-3180

Massachusetts Executive Office of Environmental Affairs
100 Cambridge St, Boston, MA 02202

Massachusetts Horticulture Society
300 Massachusetts Ave, Boston, MA 02115

Maui Project, The
PO Box 1716, Makawao, HI 96768

Max McGraw Wildlife Foundation
PO Box 9, Dundee, IL 60118

Maxwell Laboratories, Inc.
PO Box 1620, LaJolla, CA 92038

McCloud River Preserve
PO Box 409, McCloud, CA 92057

McCrone, Inc.
20 Redgeley Ave, Annapolis, MD 21401

McHenry County Conservation District
6512 Harts Rd, Ringwood, IL 60072

McKeener Environmental Learning Center
RD 3, Box 121, Sandy Lake, PA 16145

Meadowcreek Project
Fox, AR 72051

Memphis Pink Palace Museum
Coon Creek Fossil Farm, 3050 Central Ave, Memphis, TN 38111

Mendocino Woodlands Outdoor
40 Glen Dr, Mill Valley, CA 94941

Merck Forest & Farmland Center
PO Box 86, Rupert, VT 05768, (802) 394-7836

Metro District Commission
Reservations and Historical Sites, 20 Somerset St, Boston, MA 02108

Metro Education Program Agency
Gifford Educational Farm & Woodlands, 700 Camp Gifford Rd, Bellevue, NB 68005

Michael Brandman Associates, California
2530 Red Hill Ave, Santa Ana, CA 92705

Michigan Department of Natural Resources
PO Box 30028, Lansing, MI 48909

Michigan State University
W. K. Kellogg Biological Station, Hickory Corners, MI 49060

Michigan Water Resources Commission
PO Box 30028, Lansing, MI 48909

Midpeninsula Regional Open Space
201 San Antonio Circle, C-135, Mountain View, CA 94040

Midwest Research Institute
425 Volker Blvd, Kansas City, MO 64110

Midwest Sustainable Agriculture Working Group
101 S Tallman St, Walthill, NB, 68067

Midwestern Testing Labs, Inc.
54 1/2 North Main St, Fairfield, IA 52556

Miles Wildlife Sanctuary
95 West Cornwall Rd, Sharon, CT 06095

Mill Creek Park
816 Glenwood Ave, Youngstown, OH 44502

Miller Engineers & Scientists
5308 S 12th St, Sheboygan, WI
53081

Mineral Policy Center
1325 Massachusetts Ave, Ste
550, Washington, DC 20005

Minnesota Conservation Corps
500 Lafayette Rd, St. Paul, MN
55155, (612) 296-2144

*Minnesota Department of Natural
Resources*
Bureau of Human Resources,
500 Lafayette Rd, St. Paul, MN
55155, (612) 296-6494

*Minnesota Department of Natural
Resources*
Minnesota Dept of Jobs &
Training, 200 Centennial
Office Bldg, 658 Cedar St, St.
Paul, MN 55155, (612) 269-
2616

Minnesota Pollution Control Agency
520 Lafayette Rd, St. Paul, MN
55155

Minnesota Waterfowl Association
5701 Normandale Rd,
Minneapolis, MN 55424

*Mississippi Department of Wildlife,
Fisheries & Parks*
2906 Bldg, PO Box 451,
Jackson, MS 39202, (601) 362-
9212

*Mississippi Museum of Natural
Science*
111 N Jefferson St, Jackson, MS
39202, (601) 354-7303

Mississippi State University
Mississippi Cooperative
Extension, Fish and Wildlife
Resource Unit, PO Drawer BX,
Mississippi State, MS 39762

Missouri Botanical Garden
PO Box 299, St. Louis, MO
63166

*Missouri Department of
Conservation*
Human Resources Div, Box
180, Jefferson City, MO 65102,
(314) 751-4115

*Missouri Department of Natural
Resources*
PO Box 176, Jefferson City,
MO 65102

MIT Sea Grant
Massachusetts Coastal
Resources, 100 Cambridge St,
Boston, MA 02108

Mobilization for Survival, New York
45 John St, Ste 811, New York,
NY 10038

Mohican Outdoor School
21882 Shadley Valley Rd,
Danville, OH 43014, (614)
599-9753

Monmouth Conservation Foundation
Box 191, Middletown, NJ
07748

Mono Lake Committee, The
PO Box 29, Lee Vining, CA
93541

Monomet Bird Observatory
Education Dept, Box 1770,
Manomet, MA 02345

*Montana Department of Fish
Wildlife & Parks*
1420 6th Ave, Helena, MT
59620, (406) 444-4371

*Montana Department of Natural
Resources & Conservation*
1520 E 6th Ave, Helena, MT
59620

Montana Natural Heritage Progam
1515 West 6th Ave, Helena, MT
59620

Monterey Bay Aquarium
886 Cannery Row, Monterey,
CA 93940

*Monterey Bay National History
Association*
Paharo Coast Dist State Parks,
7500 Soquel Dr, Aptos, CA
95003

*Monterey County Outdoor Education
Program*
PO Box 8085, 901 Blanco
Circle, Salinas, CA 93912

*Monterey Peninsula Regional Park
District*
Box 935, Carmel Valley, CA
93924, (408) 659-5901

Montshire Museum of Science
Box 770, Norwich, VT 05055

*Morris Arboretum, University of
Pennsylvania*
9414 Meadowbrook Ave,
Philadelphia, PA 19118

Morrison-Knudson Engineers
PO Box 79, Boise, ID 83707

Morton Arboretum
Rte 53, Lisle, IL 60532

Mostardi-Platt Associates, Inc.
945 N Oaklawn Ave, Elmhurst,
IL 60126

Mote Marine Laboratory
1600 Thompson Parkway,
Sarasota, FL 34236

*Mounds View North Environmental
Learning Center*
8950 Peppard Rd, Britt, MN
55710

Mount Washington Observatory
Gorham, NH 03581, (603)
466-5564

*Mountain Trail Outdoor School,
North Carolina*
Drawer 250, Hendersonville,
NC 28793

MSE, Inc.
PO Box 3767, Butte, MT
59701, (406) 782-0463

*Multnomah Environmental
Education District*
PO Box 301039, Portland, OR
07230

Mystic Lake YMCA Camp
Box 100, Lake, MI 48632,
(517) 544-2844

Namakan West Fisheries
PO Box 2162, Los Banos, CA
93635

*National Association of Service
Conservation Corps*
1001 Connecticut NW, Ste 827,
Washington, DC 20036

National Aquarium in Baltimore
501 E Pratt St, Pier 3,
Baltimore, MD 21202

*National Association for Public
Lands*
1118 22nd St NE, 3rd fl,
Washington, DC 20037, (202)
466-2686

*National Audubon Society,
Headquarters*
700 Broadway, New York, NY
10003, (212) 979-3000

National Audubon Society, Capitol Hill Office
  666 Pennsylvania Ave SE, Washington, DC 20003, (202) 547-9009
National Audubon Society, Western Regional Office
  555 Audubon Place, Sacramento, CA 95825
National Audubon Society
  National Environmental Education Center, 613 Riversville Rd, Greenwich, CT 06831
National Coalition Against the Misuse of Pesticides
  701 E St SE, Washington, DC 20003
National Ecology Company
  16 Greenmeadow Dr, Timonium, MD 21093
National Estuarine Research
  Waquoit Bay, PO Box 92W, Waquoit, MA 02536
National Environmental Health Association
  720 S Colorado Blvd, South Tower, 970, Denver, CO 80222
National Fish & Wildlife Foundation
  1120 Connecticut Ave, Ste 900, Washington, DC 20036, (202) 797-5445
National Geographic Society
  17 & M Sts, NW, Washington, DC 20036
National Institute for Urban Wildlife
  10921 Trotting Ridge Way, Columbia, MD 21044
National Marine Fisheries Service, Headquarters
  Silver Spring Metro Center One, 1335 East West Hwy, Silver Spring, MD 20910
National Marine Fisheries Service
  Western Administrative Support Center, 7600 Sand Point Way NE, Bin C15700, Seattle, WA 98115

National Oceanic & Atmospheric Administration
  Personnel Branch 0A211, 1335 East-West Highway, Rm 3274, Silver Spring, MD 20910, (301) 713-0515
National Oceanic & Atmospheric Administration
US Department of Commerce
  Personnel Operations Branch, Rm 411, 253 Monticello Ave, Norfolk, VA 23510, (804) 441-6880
National Park Foundation
  1101 17th St NW, Washington, DC 20036
National Park Service, Headquarters
  Interior Bldg, PO Box 37127, Washington, DC 20013
National Park Service, National Capital Region
  1100 Ohio Dr SW, Washington, DC 20242
National Park Service, Mid-Atlantic
  143 S Third St, Philadelphia, PA 19106
National Park Service, North Atlantic
  15 State St, Boston, MA 02109
National Park Service, Southeast
  75 Spring St SW, Atlanta, GA 30303
National Park Service, Midwest
  1709 Jackson St, Omaha, NE 68102
National Park Service, Southwest
  Old Santa Fe Trail, PO Box 728, Santa Fe, NM 87501
National Park Service, Rocky Mountain
  PO Box 25287, Denver, CO 80225
National Park Service, Western
  600 Harrison St, Ste 600, San Francisco, CA 94107
National Park Service, Pacific Northwest
  83 S King St, Ste 212, Seattle, WA 98104
National Park Service, Alaska
  2525 Gambell St, Anchorage, AK 99503

National Parks & Conservation Association
  1776 Massachusetts Ave, NW, Washington, DC 20036
National Recreation & Park Association
  Personnel, 2775 S Quincy St, Ste 300, Arlington, VA 22206
National Research Council, National Academy of Sciences
  2101 Constitution Ave NW, Washington, DC 20418.
National Resource Consulting Service
  31-A Fayette St, Concord, NH 03301
National Science Teachers Association
  1742 Connecticut Ave NW, Washington, DC 20009
National Toxics Campaign Fund
  1168 Commonwealth Ave, Boston, MA 02134
National Toxics Campaign Fund
  825 West 187th St, #3A, New York, NY 10033, (212) 795-7654
National Trust for Historic Preservation
  1785 Massachusetts Ave NW, Washington, DC 20036
National Wildflower Research Center
  2600 FM 973 North, Austin, TX 78725
National Wildlife Action
  855 Broadway, Boulder, CO 80302, (303) 440-5850
National Wildlife Federation
  Human Resources Div, 8925 Leesburg Pike, Vienna, VA 22184, (703) 790-4522
National Wildlife Federation
  Youth Programs, 1400 16th St NW, Washington, DC 20036, (800) 245-5484
National Zoological Park
  Washington, DC 20008
Natural Areas Association
  8 Henson Place, Champaign, IL 61820

Natural Energy Trade Association
    1025 Thomas Jefferson, Ste
    410, Washington, DC 20007,
    (202) 342-7200
Natural Land Institute
    320 S Third St, Rockford, IL
    61104
Natural Lands Trust
    1031 Palmers Mills Rd, Media,
    PA 19063
Natural Resource Council of Maine
    271 State St, Augusta, ME
    04330
Natural Resources Defense Council
    40 W 20th St, New York, NY
    10011, (212) 727-2700
Natural Resources Defense Council
    1350 New York Ave, Ste 300,
    Washington, DC 20005
Natural Resources Defense Council
    71 Stevenson St, Ste 1825, San
    Francisco, CA 94105
Natural Resources Trust of Easton
    Box 187, South Easton, MA
    02375, (508) 238-6049
*Natural Science & Engineering
Research Council of Canada*
    200 Kent St, Ottawa, Ont.,
    Canada
Naturalist at Large
    Box 3517, Ventura, CA 93006,
    (805) 624-2692
Naturalist at Large
    3435 Ocean Park Blvd, Ste
    201-D, Santa Monica, CA
    90405, (213) 472-5988
Nature Center of Charlestown
    Box 82, Devault, PA 19432,
    (215) 935-9777
*Nature Center of Lee County, Junior
Museum and Planetarium*
    Box 06023, Ft. Meyers, FL
    33906
*Nature Conservancy, The, National
Headquarters*
    Human Resources, 1815 N
    Lynn St, Arlington, VA 22209,
    (703) 247-3721 (jobs listing
    hot line)
*Nature Conservancy, The, Eastern
Regional Office*
    294 Washington St, Rm 740,
    Boston, MA 02108

*Nature Conservancy, The, Midwest
Regional Office*
    1313 5th St SE, Minneapolis,
    MN 55414
*Nature Conservancy, The, Southeast
Regional Office*
    PO Box 270, Chapel Hill, NC
    27514
*Nature Conservancy, The, California
Office*
    785 Market St, San Francisco,
    CA 94013
*Nature Conservancy, The, Western
Regional Office*
    1244 Pine St, Boulder, CO
    80302
*Nature Conservancy, The, Coastal
Programs*
    250 Lawrence Hill Rd, Cold
    Spring Harbor, NY 11724
Nature Conservancy, The
    107 Cienega St, Santa Fe, NM
    87501
Nature Discovery Center
    Friends of Bellaire Parks, 4520
    Oleander, Bellaire, TX 77401
Nature Sanctuary
    407 West La Frenz, Liberty,
    MO 64068
Navajo Nation, The
    Archaeology Dept, PO Box
    689, Window Rock, AZ 86515
Navajo Natural Heritage Program
    Box 1480, Window Rock, AZ
    86515
Nawaka Outdoor School
    PO Box 308, Yorba Linda, CA
    92686
*Nebraska Department of
Environmental Quality*
    Ste 400, 1200 N St, PO Box
    98922, Lincoln, NE 68509
*Nebraska Game and Parks
Commission*
    2200 N 33rd St, PO Box
    30370, Lincoln, NE 68503
*Nebraska Sustainable Agriculture
Society*
    PO Box 736, Hartingham, NE
    68738
Neighborhood Gardens Association
    325 Chestnut St, Ste 411,
    Philadelphia, PA 19106

Neighborhood Technology
    2125 West North Ave, Chicago,
    IL 60647
NET Environmental Testing
    12 Oak Park, Bedford, MA
    01730
*Network Earth/Turner Broadcasting
System*
    One CNN Center, Box 105366,
    Atlanta, GA 30303
Nevada Cooperative Extension
    PO Box 11130, Reno, NV
    89520
*Nevada Department of Conservation
& Natural Resources*
    123 W Nye Lane, Carson City,
    NV 89710
Nevada Department of Wildlife
    Box 10678, Reno, NV 89520
Nevada Division of Forestry
    State Dept of Personnel,
    Carson City, NV 89710
New Alchemy Institute
    237 Hatchville Rd, East
    Falmouth, MA 02536
New Canaan Nature Center
    144 Oenoke Ridge, New
    Canaan, CT 06840, (203) 966-
    9577
New England Aquarium
    Central Wharf, Boston, MA
    02110
New England Wildflower Society
    180 Hemenway Rd,
    Framingham, MA 01701, (508)
    877-7630
*New Hampshire Fish & Game
Department*
    State House Annex, Rm 1,
    Concord, NH 03301
*New Hampshire Department of
Environmental Services*
    6 Hazen Dr, PO Box 95,
    Concord, NH 03302
*New Hampshire Department of
Resources, Div of Forests*
    PO Box 856, 172 Pembroke
    Rd, Concord NH 03302.
*New Hampshire Department of
Resources, Div of Parks*
    PO Box 856, 172 Pembroke
    Rd, Concord, NH 03302

New Jersey Department of
Environmental Protection & Energy
    401 E State St, Trenton, NJ
    08625, (609) 292-2885
New Jersey Environmental
Federation
    46 Bayard St, New Brunswick,
    NJ 08901
New Jersey School of Conservation
    RD 2, Box 272, Branchville, NJ
    07826, (201) 948-4646
New Mexico Natural Heritage
Program
    University of New Mexico,
    2500 Yale Blvd SE, Ste 100,
    Albuquerque, NM 87131,
    (505) 277-1991
New Mexico Natural Resources
Department
    2040 Pacheco St, Santa Fe, NM
    87505
New Mexico Department of Game &
Fish
    PO Box 25112, Santa Fe, NM
    87504
New Mexico Environmental
Department
    1190 Saint Francis Dr, PO Box
    26110, Santa Fe, NM 87502
New Pond Education Center
    101 Marchant Rd, West
    Redding, CT 06896, (203)
    938-2117
New York Botanical Garden
    200th St & Southern Blvd,
    Bronx, NY 10458
New York City Department of
Environmental Protection
    2444 Municipal Bldg, 1 Centre
    St, New York, NY 10007
New York City Parks & Recreation
    Natural Resources Group,
    1234 5th Ave, Rm 233, New
    York, NY 10029, (212) 360-
    2774
New York Department of Civil
Service
    New York State Office Bldg
    Campus, Albany, NY 12239
New York Department of
Environmental Conservation
    50 Wolf Rd, Albany, NY 12233

New York Department of
Environmental Protection
    Natural Resources Unit, 59-17
    Junction Blvd, 19th fl,
    Elmhurst, NY 11373
New York Natural Heritage Program
    Wildlife Resource Center,
    Rensselaerville, NY 12147
New York State Office of Parks,
Recreation, & Historic Preservation
    Empire State Plaza, Albany, NY
    12238, (518) 474-4492
New York-New Jersey Trail
Conference
    232 Madison Ave, Rm 908,
    New York, NY 10016
Newfound Harbor Marine Institute
    Rte 3, Box 170 SP, Big Pine
    Key, FL 33043
Newfound Lake Region Association
    RD 1, Box 826, Bristol, NH
    03222
Nez Perce Tribe
    PO Box 365, Lapwai, ID 83540
Nixon, Hargrave, Duans & Doyle
    Lincoln First Town, PO Box
    1051, Rochester, NY 14603
Nizhoni Institute
    Academy for Ecology & Energy,
    1304 Old Pecos Trail, Santa Fe,
    NM 87501
Norcross Wildlife Foundation
    Personnel Dept, Empire State
    Bldg, Rm 1322, New York, NY
    10118
Norman Bird Sanctuary
    583 Third Beach Rd,
    Middletown, RI 02840
Normandeau Associates
    PO Box 1393, Aiken, SC
    29802
North American Assoc for
Environmental Education
    PO Box 400, Troy, OH 45373
North American Environmental
Service Corp
    PO Box 66, Buffalo, NY 14207
North Carolina Botanical Garden
    3375 Totten Center, Chapel
    Hill, NC 27599

North Carolina Department of
Environmental, Health & Natural
Resources
    PO Box 27687, Raleigh, NC
    27611
North Carolina Outward Bound
    121 North Sterly St,
    Morganton, NC 28655
North Carolina State Wildlife
Resources Commission
    Archdale Bldg, 512 N Salisbury
    St, Raleigh, NC 27604
North Carolina Wildlife Federation
    Box 10626, Raleigh, NC
    27605, (919) 833-1923
North Cascades Institute
    2105 Highway 20, Sedro
    Woolley, WA 98284
North Central Experiment Station
    1816 Highway 169 East, Grand
    Rapids, MN 55744
North Country Trail Association
    1857 Torquay Ave, Royal Oak,
    MI 48073
North Dakota Parks & Recreation
    1424 W Century Ave, Ste 202,
    Bismark, ND 58501
North Dakota Department of Fish &
Game
    100 N Bismarck, Bismarck, ND
    58501
North Dakota Institute for Ecological
Studies
    PO Box 7110, Univ of North
    Dakota, Grand Forks, ND
    58202
North Kisap Marine Science
    17771 Fjord Dr NE, Poulsbo,
    WA 98370
North Prairie Wildlife Reservation
    Personnel Dept, Box 2096,
    Jamestown, ND 58402
North Woods Resource Center
    3788 North Arm Rd, Ely, MN
    55761
North Woods Resource Center
YMCA, Widjiwagan
    1761 University Ave W, St.
    Paul, MN 55104, (612) 645-
    6605

Northern Alaska Environmental Center
218 Drway, Fairbanks, AK 99701, (907) 452-5021

Northern California Area Preserves
6500 Desmond Rd, Galt, CA 95632

Northern Illinois University
Dept of Biological Sciences, DeKalb, IL 60115

Northern Illinois University
OCR Outdoors, Office of Campus Recreation, DeKalb, IL 60115, (815) 753-9418

Northern Lights Institute
Box 8084, Missoula, MT 59807

Northern Plains Resource Council
419 Stapleton Bldg, Billings, MT 59101, (406) 248-1154

Northern Rockies Action Group
9 Placer St, Helena, MT 59601

Northland College
Sigurd Olson Environmental Institute, Ashland, WI 54806

Northwest Enviroservice, Inc.
1700 Airport Way S, Seattle, WA 98124

Northwest Park Nature Center
145 Lang Rd, Windsor, CT 06095

Northwest Renewable
1133 Detor Horton Bldg, 710 2nd Lawn, Seattle, WA 98104

Northwest Rivers Council
1731 Westlake Ave N, Ste 202, Seattle, WA 85109

Northwest Youth Corps
5120 Franklin Blvd, #7A, Eugene, OR 97403, (503) 746-8653

Northwest Voyagers
PO Box 373C, Lucile, ID 83542

NSI Technology Services Corp
Human Resources Dept, 200 SW 35th St, Corvallis, OR 97333

NSI Technology Services Corp
Human Resources Dept, c/o US EPA, ESAT Region VI, 10625 Fallstone Rd, Houston, TX 77099

NSI Technology Services Corp
Human Resources Dept, PO Box 12313, 2 Triangle Dr, Research Triangle Park, NC 27709

Oak Ridge Research Institute
113 Union Valley Rd, Oak Ridge, TN 37830

O'Brien & Gere Engineers and OBG Labs
5000 Brittonfield Pkwy, Syracuse, NY 13221

Oceanic Society
Ft. Mason Center, Bldg E, San Francisco, CA 94123

O'Farrell Youth Center
PO Box 306, Woodstock, MD 21163

Ogden Environmental & Energy Services
3211 Jermantown Rd, Fairfax, VA 22030

Oglebay Institute
A. B. Brooks Nature Center, Wheeling, WV 26003, (304) 242-6855

Ohio Center of Science & Industry
280 East Broad St, Columbus, OH 43215

Ohio Environmental Protection Agency
PO Box 1049, 1800 WaterMark Dr, Columbus, OH 43266

Ohio Department of Natural Resources
Fountain Square, Columbus, OH 43224, (614) 265-6565

Ohio Wesleyan University
Dept of Biology, Association of Field Ornithologists, Inc., Delaware, OH 43015, (614) 369-4431

OHM Corporation
Windsor Industrial Park, Windsor, NJ 08561

OHM Corporation
1369 Spears Rd, Houston, TX 77067

Oklahoma Department of Wildlife Conservation
1801 N Lincoln, PO Box 53465, Oklahoma City, OK 73152, (405) 521-3851

Oklahoma Environmental Quality Offices, State Department of Health
1000 Northeast Tenth St, PO Box 53551, Oklahoma City, OK 731252, (405) 271-7337

Oklahoma Forestry Division
Dept of Agriculture, PO Box 297, Wilburton, OK 74578

Oklahoma National Heritage
Sutton Hall 303, 625 Elm St, Norman, OK 73139

Oklahoma State Parks, Oklahoma Tourism & Recreation Department
2401 N Lincoln, Ste 500, Oklahoma City, OK 73105

Old Mill Farm School
PO Box 463, Mendocino, CA 95460

Olympic Park Institute
HC 62, Box 9T, Port Angeles, WA 98362

Open Lands Project
220 S State St, Ste 1880, Chicago, IL 60604

Oracle Center for Environmental Education
Environmental Education Exchange, 3661 N Campbell Ave, Tucson, AZ 85719, (602) 888-5111

Orange County Marine Institute
PO Box 68, Dana Point, CA 92609

Oregon Department of Environmental Quality
811 SW Sixth Ave, Portland, OR 97204, (503) 229-5696.

Oregon Department of Fish & Wildlife
Box 59, Portland, OR 97207, (503) 229-5400.

Oregon Department of Fish & Wildlife
Personnel Center, 775 Court St NE, Salem, OR 97309

Oregon Department of Forestry
2600 State St, Salem, OR 97310, (503) 378-2560

Oregon Museum of Science & Industry
Cascade Science School, Hancock Field Station, Fossil, OR 97830, (503) 322-1771

Oregon Museum of Science & Industry
    Human Resources, 4015 SW Canyon Rd, Portland, OR 97221, (503) 222-2828
Oregon Natural Resources Council
    522 SW 5th St, Ste 1050, Portland, OR 97204
Orenda Wildlife Trust
    PO Box 669, West Barnstable, MA 02668, (508) 362-4798
Orinda Union School District
    Coast Environment Arts Camp, 8 Altorinda Rd, Orinda, CA 94563
Otter Lake Conservation School
    PO Box H, Greenfield, NH 03047
Outback Expeditions
    PO Box 44, Terlinqua, TX 79852
Outdoor Education Center
    Jones Gulch, La Honda, CA 94020
Outdoor Education Program
    NEED Collaboration, Box 896, Truro, MA 02666
Outdoor Leadership Training Seminar
    Breaking Through Adventures, PO Box 200281, Denver, CO 80220
Outside Insights
    1124 Riviera St, Venice, FL 34285
Outside Insights
    1055 W College Ave, Ste 327, Santa Rosa, CA 95401
Ozark Natural Science Center
    1 Timber Trails, Rogers, AR 72756, (501) 636-9258
Pacific Crest Outward Bound
    PO Box 629, Wilsonville, OR 97070
Pacific Energy & Resources Center
    PE & R Center, Bldg 1055, Ft. Cronkite, Sausalito, CA 94965
Pacific Marine Fisheries Commission
    45 S 82nd Dr, Gladstone, OR 97027
Pacific Northwest Pollution Prevention Resource Center
    Box 2777, Seattle, WA 98111

Pacific Whale Foundation
    PO Box 1038, Makena, Maui, HI 96753
PAR Group, Illinois
    100 N Waukegan Rd, Ste 200, Lake Bluff, IL 60044, (708) 234-8309
Paradigm Ventures
    29 Azul Loop, Santa Fe, NM 87505
Parks Council, The
    457 Madison Ave, New York, NY 10022
Patrick Engineering, Inc.
    346 Taft Ave, Glen Ellyn, IL 60137
Passport for Adventure
    PO Box 1340, Salina, KS 67402
Pathfinder Outdoor Science School
    104 Garner Valley, Mountain Center, CA 92561
PDG Environmental Group, Inc.
    300 Oxford Dr, Monroeville, PA 15146
Peace Valley Nature Center
    170 Chapman Rd, Doylestown, PA 18901, (215) 345-7860
Pecos River Learning Center, Inc.
    1800 Old Pecos Trail, Santa Fe, NM 87501
Pendalouan Outdoor Resource Center
    1243 Fruitvale Rd, Montague, MI 49437
Pennoni Associates
    1600 Callow Hill Rd, Philadelphia, PA 19130
Pennsylvania Department of Environmental Resources
    16th fl, MSSOB, PO Box 2063, Harrisburg, PA 17105, (713) 783-4759
Pennsylvania Environmental Council
    225 S 15th St, Ste 506, Philadelphia, PA 19102
Pennsylvania Federation of Sportsmans Clubs
    2426 N 2nd St, Harrisburg, PA 17110

Pennsylvania State University
    Shaver's Creek Raptor Center, 203 South Henderson Bldg, University Park, PA 16802, (814) 863-2000
Pennypack Wilderness and Watershed
    2955 EdgeHill Rd, Huntington Valley, PA 19006
People's Place II, Inc.
    Outdoor Delaware, 219 South Walnut St, Milford, DE 19963, (302) 422-8011
Pesticide Degradation Laboratory
    Bldg 050, Rm 100, Beltsville, MD 20705
Philadelphia Youth Service Corps
    1617 JFK Blvd, Philadelphia, PA 19103, (215) 568-0399
Pickering Creek Environmental Center
    27370 Sharp Rd, Easton, MD 21601, (410) 822-4903
Pinchot Institute for Conservation
    PO Box 188, Milford, PA 18337
Pine Grove Child Development Center
    32 Foreside Rd, Falmouth, ME 04105
Plains Conservation Center
    3439 Uravan Way, Rm 308, Aurora, CO 80013
Pocono Environmental Education Center
    RD 2, Box 1010, Dingmans Ferry, PA 18328, (717) 828-2319
Point Fermin Outdoor School
    Los Angeles Unified School District, 450 N Grand Ave, Rm G-314, Los Angeles, CA 90012
Point Reyes Bird Observatory
    4990 Shoreline Highway, Stinson Beach, CA 94970
Pok-o-Mac Cready Outdoor Education
    289 Mountain Rd, Willsboro, NY 12996
Pollution Control Labs
    960 Remillard Ct, San Jose, CA 95122

Population Environment Balance
1325 G St NW, Ste 1003,
Washington DC 10003

Portland General Electric
121 SW Salmon St, Portland,
OR 97204

Potomac Appalachian Trail Club
118 Park St SE, Vienna, VA
22180, (703) 242-0963

Powder River Basin Research Center
23 North Scott, Sheridan, WY
82801

Powder River Resource Council
PO Box 1178, Douglas, WY
82801

Pratt Center
163 Paper Mill Rd, New
Milford, CT 06776

Pratt Institute
200 Willoughby Ave,
Brooklyn, NY 11205

PRC Environmental Management
233 N Michigan Ave, Chicago,
IL 60601, (312) 856-8700

Prescott College
Environment Studies, 220
Grove Ave, Prescott, AZ 86301

Princeton-Blairstown Center
158 Millbrook Rd, Blairstown,
NJ 07825

Princeton Education Center
1797 DJ Princeton University,
Princeton, NJ 08544

Project Oceanology
Avery Point, Groton, CT 06340

*Project Urban Suburban
Environments (USE)*
Box 3315, Long Branch, NJ
07740, (908) 870-6650

Public Citizen's Congress Watch
215 Pennsylvania Ave SE,
Washington, DC 20003

*Public Employees for Environmental
Responsibility*
Box 428, Eugene, OR 97440,
(503) 484-7158

Public Interest Research Groups
29 Temple Place, Boston, MA
02111

Public Museum of Grand Rapids
Grand Rapids City Hall, 300
Monroe Ave NW, Grand
Rapids, MI 49503

Public-Private Ventures
399 Market St, Philadelphia,
PA 19106

Puget Power
Bellevue, WA 98005

*Puget Sound Water Quality
Authority*
217 Pine St, Ste 1900, Seattle,
WA 98101

*Putnam-Northern Westchester
Board of Cooperative Education
Services*
Outdoor Education Program,
Yorktown Heights, NY 10598

Pyramid Environmental Systems
PO Box 548, Sparta, NJ 07871

Quebec-Labrador Foundation (QLF)
39 S Main St, Ipswich, MA
01938

Queens College
Center for Environmental
Teaching & Research, Caumsett
State Park, Huntington, NY
11743, (516) 421-3526

Rachel Carson Council
8940 Jones Mill Rd, Chevy
Chase, MD 20815

Radian Corp
PO Box 201088, Austin, TX
78720

Rails-to-Trails Conservancy
1400 16th St NW, #300,
Washington, DC 20036

Rainforest Action Network
301 Broadway, Ste A, San
Francisco, CA 94133

Rainforest Alliance
270 Lafayette St, #512, New
York, NY 10012, (210) 941-
4986

Rancho Santa Ana Botanic Garden
1500 North College,
Claremont, CA 91711

Raptor Center, The
1920 Fitch Ave, St. Paul, MN
55108

*RARE Center for Tropical
Conservation*
1529 Walnut St, Philadelphia,
PA 19102

*Red Cliff Band, Lake Superior
Chapter*
PO Box 529, Bayfield, WI
54814

REMCOR, Inc.
701 Alpha Dr, PO Box 38310,
Pittsburgh, PA 15238, (412)
963-1106

Renew America
1400 16th St NW, Ste 710,
Washington, DC 20036, (202)
232-2252

Republic Environmental Systems
101 Jessup Rd, Thorofare, NJ
08086

Resource Integration Systems
1 Salmon Brook St, Granby,
CT 06035

Resource Management Group, Inc.
Box 487, Grand Haven, MI
49417

Resources Conservation Co
3006 Northrup Way, Ste Two,
Bellevue, WA 98004

Resources for the Future
1616 P St NW, Washington,
DC 20036

Rettew Associates, Inc.
5010 Ritter Rd, Ste 102,
Mechanicsburg, PA 17055

*Rhode Island Department of
Environmental Mgmt*
9 Hayes St, Providence, RI
02908

RI Lung
10 Abbott Park Place,
Providence, RI 02903

Ridges Sanctuary, Inc.
PO Box 113, Baileys Harbor,
WI 54202

Riedel Environmental Services
4611 N Channel Ave, Portland,
OR 97217

River Bend Nature Center
Box 265, Faribault, MN 55021,
(507) 332-7151

River Network
Box 8787, Portland, OR
97207, (800) 423-6747

*Riverbend Environmental Education
Center*
Box 2, Gladwyne, PA 19035,
(215) 527-5234

Riveredge Nature Center
Box 26, Newburg, WI 53060, (414) 375-2715

RMC Environmental Services
1921 River Rd, Box 10, Drumore, PA 17518, (717) 548-2121

Roaring Fork Land Conservancy
210-PP A.B.C., Aspen, CO 81611

Rocky Mountain Elk Foundation
3625 Soderberg Dr, Ft. Collins, CO 80526

Rockywold-Deephaven Camps
Box B, Holderness, NH 03245, (603) 968-3313

Rodale Press
33 East Minor St, Emmaus, PA 18098

Rodale Research Center
RD 1 Box 323, Kutztown, PA 19530

Roger Tory Peterson Institute of Natural History
110 Marvin Parkway, Jamestown, NY 14701

Roger Williams Park Zoo
Elmwood Ave, Providence, RI 02905

Rogers Environmental Education Center
PO Box 716, Sherburne, NY 13460

Rollins Environmental Services
One Rollins Plaza, Wilmington, DE 19899

Roy F. Weston, Inc.
One Weston Way, West Chester, PA 19380

Ruffner Mountain Nature Center
1214 South 81st St, Birmingham, AL 35206, (205) 833-8112

Rutgers Shellfish Research Laboratory
PO Box 687, Port Norris, NJ 08349

Sacramento Science Center
City of Sacramento, 801 9th St, Ste 101, Sacramento, CA 95814

Safe Energy Communication Council
1717 Massachusetts Ave, Ste LL215, Washington, DC 20036

Safe Environments
2512 9th St, Rm 17, Berkeley, CA 94710

Sagamore Institutes
Box 106, Sagamore Rd, Raquette Lake, NY 13436

SAIC
1710 Goodridge Dr, McLean VA 22102, (703) 821-4300.

Saltwater, Inc.
540 L St, Ste 202, Anchorage, AK 99501, (907) 258-5999

San Benito County Parks
440 5th St, Rm 206, Hollister, CA 95023

San Francisco Conservation Corps
Bldg 111, San Francisco, CA 94123

San Francisco Recycling Program
City Hall, Rm 271, San Francisco, CA 94102

San Francisco State University
Wildlands Studies, 3 Mosswood Circle, Cazadero, CA 94521, (707) 632-5665

San Joaquin River Parkway
668 W Shaw Ave, Ste F-174, Fresno, CA 93704

San Mateo Outdoor Education
La Honda, CA 94020, (415) 747-0414

Sandyland Sanctuary
PO Box 909, Silsbee, TX 77656

Sanibel-Captiva Conservation Foundation
Box 839, Sanibel, FL 33957, (813) 472-2329

Santa Cruz County
126B Rancho Del Mar, Aptos, CA 95003

Santa Cruz County Office of Education
809-H Bay Ave, Capitola, CA 95010

Santa Fe Mountain Center
Rte 4, Box 34-C, Santa Fe, NM 87501

Sarett Nature Center
2300 Benton Center Rd, Benton Harbor, MI 49022

Save America's Forests
4 Library Court SE, Washington, DC 20003

Save the Bay
434 Smith St, Providence, RI 02908

Save the Harbor-Save the Bay
PO Box 109, Boston, MA 02133

Save the Manatee Club
500 N Maitland Ave, Maitland, FL 32751

Save-the-Redwoods League
114 Sansome St, San Francisco, CA 94104

Scenic America
21 Dupont Circle, NW, Washington, DC 20036

Scenic Hudson, Inc.
9 Vassar St, Poughkeepsie, NY 12601

Scherman-Hoffman Sanctuaries
PO Box 693, Bernardsville, NJ 07924

Schlitz Audubon Center
1111 East Brown Deer Rd, Milwaukee, WI 53217

Schmid & Company, Inc.
1201 Cedar Grove Rd, Media, PA 19063

School for Field Studies
16 Broadway, Beverly, MA 01915

School for International Learning
Brattleboro, VT 05301

School of Environmental Education
Rte 2, Box 3197, Navasota, TX 77868

Schooner, Inc.
60 South Water St, New Haven, CT 06519

SCICON
PO Box 339, Springville, CA 93265, (209) 539-2642

Science Center of New Hampshire
PO Box 173, Holderness, NH 03245

Science Museum of Minnesota
St. Croix Field Research Center, 30 East Tenth St, St. Paul, MN 55101

Scientific Resources, Inc.
11830 SW Kerr Parkway, Ste 375, Lake Oswego, OR 97035

Scientists' Institute for Public Information
355 Lexington Ave, New York, NY 10017

Sea Critters, Inc.
50 Sea Critters Lane, Key Largo, FL 33037

Sea Shepherd Conservation Society
1314 Second St, Santa Monica, CA 90401

Seacamp, Florida
Rte 3, Box 170, Big Pine Key, FL 33043, (305) 872-2331

Seacoast Science Center
Box 674, Rye, NH 03870

Seafarers Expeditions
PO Box 691, Bangor, ME 04401

Seatlantic, Inc.
PO Box 1004, Katy, TX 77492

Seattle Aquarium
Waterfront Park, Pier 59, Seattle, WA 98101

SEC Donohue, Inc.
PO Box 24000, Greenville, SC 29616

SEECO Environmental Services, Inc.
7350 Duvan Dr, Tinley Park, IL 60477

Selby Botanical Gardens
811 South Palm Ave, Sarasota, FL 34230

Seven Hawks Wilderness Program
Rte 3, Box 87A, Waverly, TN 37185

Sheridan School Mountain Campus
Box 436, RFD 4, Luray, VA 22835, (703) 743-7655

Sherkin Island Marine Station
Sherkin Island, County Cork, Ireland, 028

Sierra Club, National Headquarters
730 Polk St, San Francisco, CA 94109, (415) 776-2211

Sierra Club, Washington, D.C. Office
408 C St NE, Washington, DC 20003

Sierra Club Legal Defense Fund, National Headquarters
180 Montgomery St, Ste 1400, San Francisco, CA 94104

Siemens Solar Industries
4650 Adohr Lane, Camarillo, CA 93010

Silver Cloud Expeditions
PO Box 1006-B, Salmon, ID 83467

Sky Ranch
Rte 1, Box 60, Van, TX 75790, (903) 569-3482

Slide Ranch
Muir Beach, CA 94965

Smithsonian Institution
1000 Jefferson Dr SW, Washington, DC 20560

Society for Protection of New Hampshire Forests
54 Portsmouth St, Concord, NH 03301

Society of American Foresters
5400 Grosvenor Lane, Bethesda, MD 20814, (301) 897-8720

Soil & Water Conservation Society
7515 NE Ankeny Rd, Ankeny, IA 50021

Solarex Corp
630 Solarex Ct, Frederick, MD 21701

Sound Experience
Box 2098, Poulsbo, WA 98370

Soundwaters
4 Yacht Haven Marine Center, Washington Blvd, Stamford, CT 06902, (203) 323-1978

South Coast Air Quality Management
9150 Flair Dr, El Monte, CA 91731

South Community Land Trust
PO Box 23468, Providence, RI 02903

South Carolina Department of Health & Environmental Control
2600 Bull St, Columbia, SC 29201

South Carolina Department of Parks, Recreation & Tourism
1205 Pendleton St, Columbia, SC 29201

South Carolina Division of Natural Resources
1205 Pendleton St, Office of the Governor, Columbia, SC 29201

South Carolina Forestry Commission
Box 21707, Columbia, SC 29221

South Carolina Heritage Trust
PO Box 167, Columbia, SC 29202

South Dakota Department of Environment & Natural Resources
523 E Capitol, Pierre, SD 57501

South Dakota Department of Fish & Parks
Anderson Blvd, Pierre, SD 57501

South Shore Nature Science Center
Box 429, Norwell, MA 02061

South Shore YMCA Camps
Stowe Rd, RD 2, Sandwich, MA 02563

South Slough National Estuarine Resource Reserve
Box 5417, Charleston, OR 97420, (503) 888-5558

Southeast Alaska Conservation Council
PO Box 021692, Juneau, AK 99802, (907) 586-6942

Southern Appalachian Highlands Conservancy
3213 Montlake Dr, Knoxville, TN 37920, (615) 434-2555

Southern Environmental Law Center
210 W Main St, Ste 14, Charlottesville, VA 22901

Southern Illinois University
Touch of Nature Environmental Center, Carbondale, IL 62901, (618) 453-1121

Southwest Florida Water Management District
2379 Broad St, Brooksville, FL 34609

Southwind Adventures, Inc.
1861 Camino Lumbre, Ste 1, Santa Fe, NM 87505

Special Interest Tours
134 W 26th St, New York, NY 10001

Sport Fishing Institute
1010 Massachusetts Ave NW, Ste 320, Washington, DC 20001, (202) 898-0770

St. Croix Chippewa Reservation
PO Box 287, Hertel, WI 54845

St. Lawrence-Eastern Ontario Commission
Dulles State Office Bldg, 317 Washington St, Watertown, NY 13601

St. Louis Science Center
Personnel E/E, 5050 Oakland Ave, St. Louis, MO 63110

Stamford Marina
Long Island Sound Task Force, 185 Magee Ave, Stamford, CT 06902

Stamford Marine Center
185 Magee Ave, Stamford, CT 06902

Standard Laboratories, Inc.
4815 McCorkle Ave S, Charleston, WV 25304

Stanley M. Hunts Associates
354 Pequot Ave, Southport, CT 06490

State University of New York College at Cortland
Coalition for Education in the Outdoors, Box 2000, Cortland, NY 13045

Stone Environmental Schools
PO Box 165, Ocean Park, ME 04063, (207) 934-4064

Stony Brook Nature Center
North St, Norfolk, MA 02056, (508) 528-3140

Storm King School
Cornwall-on-Hudson, NY 12520

Storm Mountain Center
HRC 33, Box 1701, Rapid City, SD 57001

Strybing Arboretum
9th Ave at Lincoln Way, San Francisco, CA 94122

Student Conservation Association
Box 550, Charlestown, NH 03603, (603) 543-1700

Student Environmental Action Coalition
Box 1168, Chapel Hill, NC 27514, (919) 967-4600

Student Pugwash
1638 R St NW, Ste 32, Washington, DC 20009

Successful Communities, Florida
1000 Friends of Florida, PO Box 5948, Tallahassee, FL 32314

Sunrise Lake Outdoor Education Center
7 North 749 Rte 59, Bartlett, IL 60103

SuperCamp
1725 South Hill St, Oceanside, CA 92054, (800) 527-5321

Swanson Environmental, Inc.
24156 Haggerty Rd, Farmington Hills, MI 48335

Taconic Outdoor Education Center
12 Dennytown Rd, Cold Spring, NY 10516

Tahoe Regional Planning Agency
PO Box 1030, Zephyr Cove, NV 89448

Tanglewood 4-H Camp
375 Main St, Rockland, ME 04841, (207) 594-2104

Technology Information Project
PO Box 1371, Rapid City, SD 57009

Tenafly Nature Center
313 Hudson Ave, Tenafly, NJ 07670, (201) 568-6093

Tennessee Department of Conservation
Div of Ecological Services, 701 Broadway, Nashville, TN 37219

Tennessee Department of Environment & Conservation
401 Church St, Nashville, TN 37243

Tennessee Mountain Management
12541 Chapman Highway, Seymour, TN 37865

Tennessee Wildlife Resources Agency
PO Box 40747, Ellington Agricultural Center, Nashville, TN 37204

Terra Verde Trading Company
13-17 Laight St, 5th fl, New York, NY 10013, (212) 941-4658

Teton Science School
Box 68, Kelly, WY 83011

Tetra-Tech, Inc.
11820 Northrup Way, Ste 100, Bellevue, WA 98005

Tetra-Tech, Inc.
670 N Rosemead Blvd, Pasadena, CA 91107, (818) 449-6400

Texas A & M University
Dept of Forest Science, College Station, TX 77843

Texas Forest Service
College Station, TX 77843

Texas Parks & Wildlife Department
4200 Smith School Rd, Austin, TX 78744, (512) 389-4809

Texas Tech University
American Society of Mammalogists, The Museum, Lubbock, TX 79409, (806) 742-2442

Thompson Island Outward Bound
Box 127, Boston, MA 02127

Thorne Ecological Institute
5398 Manhattan Circle, Boulder, CO 80303

Tighe & Bond, Inc.
53 Southampton Rd, Westfield, MA 01085.

Tocikon Corporation
1860 Old Okeechobee Rd, Ste 401, West Palm Beach, FL 34409

Toledo Zoo
2700 Broadway, Toledo, OH 43609

Tuolumne County
25 Green St, Sonora, CA, 95370

Toxicon Environmental Science
535 E Indiantown Rd, 1st fl, Jupiter, FL 33477

Toxikon Corporation
225 Wildwood Ave, Woburn, MA 01801

Trail Blazers, Inc.
275 7th Ave, 12th fl, New York, NY 10001

Trailmark Outdoor Adventures
155 Schuyler Rd, Nyack, NY
10960

TRC Environmental Corporation
Boott Mills South, Foot of
John St, Lowell, MA 01852,
(508) 970-5600

TreePeople
12061 Mulholland Dr, Beverly
Hills, CA 90201, (818) 753-
4620

Trees New York
NYC St Tree Consortium, 44
Worth St, New York, NY
10013, (212) 766-2863

TREK Program Project
PO Box 837, Red Bank, NJ
07701, (908) 291-7300

Tressler Care Wilderness School
PO Box 10, Boiling Springs, PA
17007, (717) 258-3168

Trinity Consultants, Inc.
12801 N Central Expressway,
Ste 1200, Dallas, TX 75243

*Tropical Research and Development,
Inc.*
519 NW 60th St, Ste D,
Gainesville, FL 23607

Trout Unlimited
501 Church St NE, Vienna, VA
22180, (703) 281-1100

Trust for Appalachian Trail Lands
PO Box 122, Norwich, VT
05055

Trust for Hidden Village, The
HVEEP Internship Program,
26870 Moody Rd, Los Altos
Hills, CA 94022

Trust for Public Land, The
Human Resources Dept, 510
1st Ave North, Ste 210,
Minneapolis, MN 55403

Trust for Public Land, The
Human Resources Div, 116
New Montgomery St, 4th fl,
San Francisco, CA 94105

Trust for Public Land, The
PO Box 2383, Santa Fe, NM
87504

Trust for Public Land, The
666 Broadway, New York, NY
10012

Trust for Public Land, The
322 Beard St, Tallahassee, FL
32303

Trustees for Alaska
725 Christensen Dr, #4,
Anchorage, AK 99501

Trustees of Reservations
Martha's Vineyard, 183
Whiting St, Hingham, MA
02043

Tucson Botanical Gardens
2150 North Alvernon Way,
Tucson, AZ 95712

*Turtle Mountain Environmental
Learning Center*
RR 1, Box 359, Bottineau, ND
58318, (701) 263-4514

Tybee Island 4-H Center
PO Box 1477, 9 Lewis Ave,
Tybee Island, GA 31328, (912)
786-5534

Tyson Research Center, RRPP, Inc.
Research Center, PO Box 193,
Eureka, MO 63025

UNCC
Venture Program, Cone
University Center, Charlotte,
NC 28223

Union of Concerned Scientists
26 Church St, Cambridge, MA
02238

*United Nations Environmental
Program*
Chief Recruitment Unit, Box
30552, Nairobi, Kenya, fax
(2542) 228890

United States Army
Fish & Wildlife Branch, Bldg
1938, Ft. Hood, TX 76544,
(817) 287-3114

*United States Army, Civil
Engineering Research Lab*
PO Box 4005, Champaign, IL
61824, (217) 398-6443

United Water Conservation District
PO Box 432, Santa Paula, CA
93060

University of Alaska, Anchorage
3890 University Lake Dr,
Anchorage, AK 99508, (907)
786-4608

University of Arkansas
Dept of Zoology, Fayetteville,
AR 72701

University of Arizona
College of Agriculture, Forbes
320, Tucson, AZ 85721

University of Buffalo
Office of the Provost, 562
Capen Hall, Buffalo, NY
14260

University of Calgary
2600 University Dr NW,
Calgary, Alta., Canada

University of California
Bldg of Environmental Studies,
Santa Cruz, CA 95064

University of California, Davis
DANR, North Region, Davis,
CA 95616, (916) 757-8621

University of California Extension
Elkus 4-H Ranch, 1500
Purisima Creek Rd, Half Moon
Bay, CA 94019

University of California, Irvine
Program in Social Ecology,
Irvine, CA 92717

University of Colorado
Center of Recreation &
Tourism Development,
Campus Box 420, Boulder, CO
80309, (303) 492-3725

*University of Colorado Outdoor
Program*
Campus Box 355, Boulder, CO
80309

University of Florida
Box 875, Noceville, FL 32588,
(904) 729-3399

University of Florida
Dept of Wildlife & Range
Sciences, 118 Newins-Ziegler
Hall, Gainesville, FL 32611

University of Georgia
Rte 4, Box 1376, Dahlonega,
GA 30533, (404) 864-2050

University of Idaho
Garton Fish & Wildlife,
Moscow, ID 83844, (208) 885-
6434

*University of Illinois at Urbana-
Champaign*
506 S Wright St, Urbana, IL
61801

University of Iowa
Warren Slebos Div of Recreation, E216 Field House, Iowa City, IA 52242

University of Kentucky
Faculty Scholars Program, 110 Maxwellton Court, Lexington, KY 40506

University of Maine Cooperative Extension Service
York County Office, Alfred, ME 04002

University of Maryland
Appalachian Environmental Laboratory, Frostburg, MD 21532

University of Michigan
4901 Evergreen Rd, Dearborn, MI 48128

University of Minnesota
Dept of Biology, Duluth, MN 55812

University of Missouri
Rm 1-35, Agricultural Bldg, Columbia, MO 65211

University of Montana
Missoula, MT 59812

University of North Carolina
Environmental Studies Program, Asheville, NC 28804

University of Nebraska at Omaha
Outdoor Venture Center, HPER 100, University of Nebraska, Omaha, NB, 68182

University of Nevada
Dept of Range, Wildlife & Forest, 100 Valley Rd, Reno, NV 89512

University of Puget Sound
School of Education, 1500 North Warner, Tacoma, WA 98416, (206) 756-3250

University of Rhode Island
Environmental Education Center
401 Victory Highway, West Greenwich, RI 02817, (401) 397-3304

University of Tennessee
Dept of Forestry, Wildlife & Fisheries, Knoxville, TN 37901, (615) 974-7126

University of Texas
Dept of Zoology, Austin, TX 78712, (512) 471-9935

University of Toledo
Dept of Biology, Toledo, OH 43606

University of Vermont
Dept of Microbiology, Given Bldg, Burlington, VT 05405

University of Virginia
Virginia Coast Long-term Ecological, Charlottesville, VA 22901

University of Washington
American Society of Limnology & Oceanography, Inc., WB-10, Seattle, WA 98195

University of Wisconsin Extension
Cooperative Extension Personnel, 619 Extension Bldg, 432 N Lake St, Madison, WI 53706, (608) 263-1945

Upham Educational Center
Upham Woods 4-H Environmental Education Center, NI94 County Trunk N, Wisconsin Dells, WI 53965, (606) 254-6461

Urban League of Flint
202 East Boulevard Dr, Flint, MI 48503

Urbana Park District
505 N Broadway, Urbana, IL 61801

U.S. Geological Survey
12201 Sunrise Valley Dr, Reston, VA 22092

USDA Soil Conservation Service
Rm 6218-S, Box 2890, Washington, DC 20013

Utah Department of Health
Environmental Division
PO Box 16700, Salt Lake City, UT 84116,

Utah State Department of Natural Resources
1636 W North Temple, Salt Lake City, UT 84116, (801) 538-7200

Utica Zoo
Steele Hill Rd, Utica, NY 13501

Vanderbilt University
SPEHP, Station 17, Nashville, TN 37232

Vanguard Communications
1835 K St NW, Washington, DC 20006

Ventana Wilderness Sanctuary
8590 Carmel Valley Rd, Carmel, CA 93923, (408) 626-8348

Venture West School of Outdoor Living
1924 Magnolia Way, Walnut Creek, CA 94595

Vermont Agency of Natural Resources
103 S Main St, Waterbury, VT 05677

Vermont Institute of Natural Science
Box 86, Woodstock, VT 05091, (802) 457-2779

Vermont Fish & Wildlife Department
103 S Main St, Waterbury, VT 05676, (802) 244-7340

Vermont Natural Resources Council
9 Bailey Ave, Montpelier, VT 05602

Vermont Youth Conservation Corps
103 S Main St, Waterbury, VT 05671, (802) 455-5045

Virginia Department of Conservation & Recreation
203 Governor St, Ste 302, Richmond, VA 23219, (804) 786-2121

Virginia Department of Game & Inland Fisheries
4010 W Broad St, Box 11104, Richmond, VA 23230, (804) 367-1000

Virginia Department of Transportation
Human Resource Office, 3975 Fair Ridge Dr, Fairfax, VA 22033

Virginia Museum of Natural History
1001 Douglas Ave, Martinsville, VA 24112

Virginia Polytechnic Institute & State University
Blacksburg, VA 24061

Virginia, Northern, Regional Park Authority
5400 Ox Rd, Fairfax Station, VA 22039, (703) 352-5900

Voyageur Outward Bound School
10900 Cedar Lake Rd, Minnetonka, MN 55343

Voyagers International
PO Box 915, Ithaca, NY 14851

Wadsworth-Alert Laboratories
4101 Shufel Dr, NW, North Canton, OH 44720

W. B. McCloud & Company
Aquatic Weed Services, 1012 West Lunt, Schaumburg, IL 60193

Wackenhut Corporation
1500 San Remo Ave, Coral Gables, FL 33146, (305) 662-7356

Wahsega 4-H Center
Rte 8, Box 1820, Dahlonega, GA 30533, (706) 864-2050

Waikiki Aquarium
2777 Kalakaua Ave, Honolulu, HI 96815, (808) 923-9741

Waquoit Bay National Estuarine Resource Reservation
Box 3092W, Waquoit, MA 02536, (508) 457-0495

Washington County Outdoor School
17705 NW Springville Rd, Portland, OR 97229

Washington Department of Ecology
Olympia, WA 98504-8711, (206) 459-6000

Washington Department of Fisheries
PO Box 43135, Olympia, WA 98504-3135, (206) 753-6600

Washington Department of Natural Resources
PO Box 47001, Olympia, WA 98504-7001, (206) 902-1000

Washington State Parks & Recreation Commission
7150 Cleanwater Lane, Olympia, WA 98504-5711, (206) 753-5760

Washington State University
Dept of Natural Resource Science, Pullman, WA 99164, (509) 335-8848

Washington Wilderness Coalition
PO Box 458187, Seattle, WA 98145

Waste Stream Technology, Inc.
302 Grote St, Buffalo, NY 14207

Waterman Conservation Education Center
Box 288, Hilton Rd, Appalachia, NY 13732

WaterWatch
921 SW Morrison, Ste 438, Portland, OR 97205

Web of Life Outdoor Education
Box 530, Carver, MA 02330

Wehran Envirotech, Inc.
666 E Main St, Middletown, NY 10940

Weiss Ecology Center
150 Snake Den Rd, Ringwood, NJ 07456

Westchester County Environmental Management
Div of Environmental Planning, 427 Michaelian Office Bldg, White Plains, NY 10601

Western Colorado Congress
Box 472, Montrose, CO 81402, (303) 249-1978

Western Michigan Environmental Action Council
1432 Wealthy St SE, Grand Rapids, MI 49506

Western Michigan University
Dept of Science Studies, Kalamazoo, MI 49008

Western Wisconsin Technical College
304 North 6th St, LaCrosse, WI 54601

Westinghouse Remediation Services
675 Park North Blvd, Clarkston, GA 30021

Westmoreland Sanctuary
RD 2, Chestnut Ridge Rd, Mt. Kisco, NY 10549

Westport Environmental Systems
251 Forge Rd, Westport, MA 02790

Westvaco
Timberlands Div, Rte 3, Box 945, Ridgeville, SC 29472

West Virginia Division of Natural Resources
1900 Kanawha Blvd East, Charleston, WV 25305, (304) 558-2754

Wetlands
Pines & Prairie, Rte 2, Box 45A, Warren, MN 56762

Whiskeytown Environmental School
Shasta County Schools, 1644 Magnolia Ave, Redding, CA 96001

White House Fellowship Program
712 Jackson Place NW, Washington, DC 20503

White Mountain School
West Farm Rd, Littleton, NH 03561

Whitefish Point Bird Observatory
HC 48, Box 115, Paradise, MI 49768, (906) 492-3596

Whole Earth Landscaping
Box 812022, Wellesley, MA 02181, (508) 655-0722

Wild Canid Survival & Research Center
Box 760, Eureka, MO 63025

Wild Horse Organized Assistance
PO Box 555, Reno, NV 89504

Wilderness Aware Rafting
Internship Program for Alaska, PO Box 1550, Buena Vista, CO 81211, (719) 395-2112

Wilderness Center, The
Box 202, Wilmont, OH 44689

Wilderness Challenge School
119-B Tilden Ave, Chesapeake, VA 23320

Wilderness Conquest, Inc.
PO Box 50431, Idaho Falls, ID 83405, (208) 522-1080

Wilderness Education Association
20 Winona Ave, Box 89, Saranac Lake, NY 12983

Wilderness Idaho
911 Preacher Creek Rd, Shoshone, ID 83352, (208) 886-2565

Wilderness Program, The
Rte 3, Box 6715, Twin Falls, ID 83301

Wilderness School
PO Box 2241, Goshen, CT 06756

Wilderness Society, The
900 17th St, Washington, DC 20006, (202) 833-2300

Wilderness Southeast
711 Sandtown Rd, Savannah, GA 31410

Wilderness Ventures
PO Box 2768-JA, Jackson, WY 83001

Wildlife Associates
PO Box 982, Pacifica, CA 94044

Wildlife Center of Virginia
PO Box 98, Weyers Cave, VA 22980

Wildlife Discovery Program
Houston Zoological Gardens, 1513 North Macgregor Way, Houston, TX 77030

Wildlife Federation of Alaska
PO Box 103782, Anchorage, AK 99510

Wildlife Habitat Enhancement Council
1010 Wayne Ave, Ste 1240, Silver Spring, MD 20910, (301) 588-8994

Wildlife Information Center
629 Green St, Allentown, PA 18049

Wildlife International Ltd
305 Commerce Dr, Easton, MD 21601, (301) 822-8600

Wildlife Management Institute
Ste 801, 1101 14th St, NW, Washington, DC 20005, (202) 408-5059

Wildlife Prairie Park
3826 N Taylor Rd, RR 2, Box 50, Peoria, IL 61615

Wildlife Research Center
317 West Prospect, Ft. Collins, CO 80520

Wildlife Society, The
5410 Grosvenor Lane, Bethesda, MD 20814, (301) 897-9770

Wildwood Outdoor Education Center
Rte 1, Box 76, La Cygne, KS, 66040, (913) 757-4500

Williams College
Center for Environmental Studies, Kellogg House, Williamstown, MA 01267

Wilmington Garden Center
503 Market St, Wilmington, DE 19801

Wilmington Trust Company
Wilmington, DE 19890

Winston Foundation for World Peace
1875 Connecticut Ave NW, Ste 7101, Washington, DC 20009

Wisconsin Conservation Corps
30 W Mifflin, Ste 406, Madison, WI 53703

Wisconsin Department of Natural Resources
Rm EW, Box 7921, Madison, WI 53707, (608) 266-5898

Wisconsin Environmental Decade
14 W Mifflin St, Ste 5, Madison, WI 53703

Wohelo
The Luther Gulick Camps, Box 39, South Casco, MI 04077, (207) 655-4739

Wolf Ridge Environmental Learning Center
230 Cranberry Rd, Finland, MN 55603, (218) 353-7414

Wolfcreek Wilderness School
Rte 1, Box 1190, Blairsville, GA 30512, (404) 745-5553

Wolverine Camps
Wolverine, MI 49799

Woodland Altars Outdoor Education Center
33200 State Rd 41, Peebles, OH 45660

Woodland Management
1600 Ponder Point, Sandpoint, ID 83864, (208) 263-2914

Woodlands Mountain Institute
Box 907, Franklin, WV 26807, (304) 358-2401

Woods Hole Oceanographic Institute
Box 54P, Woods Hole, MA 02543

Woodswomen, Inc.
25 W Diamond Lake Rd, Minneapolis, MN 55419, (612) 822-3809

World Environment Center
419 Park Ave South, Ste 1800, New York, NY 10016, (212) 683-5053

World Fellowship Center
Box 2880-J, Conway, NH 03818, (603) 356-5208

World Peace University
Dept of Integrated Ecology, Box 10869, Eugene, OR 97440

World Resources Institute
1709 New York Ave NW, Ste 400, Washington, DC 20006

World Society for the Protection of Animals
29 Perkins St, PO Box 190, Boston, MA 02130

World Vision
919 West Huntington Dr, Monrovia, CA 91016

World Wildlife Fund
Human Resources Dept, 148J-S 24th St NW, Washington, DC 20037, (202) 861-8350 (jobs hot line)

Worldwatch Institute
1776 Massachusetts Ave NW, Washington, DC 20036

Wyoming Department of Environmental Quality
122 W 25th St, 4th fl, Herschler Bldg, Cheyenne, WY 82002

Wyoming Bureau of Land Management
Box 1828, Cheyenne, WY 82003

Wyoming Game & Fish Department
2800 Pheasant Dr, Casper, WY 82604, (307) 234-9185

Wyoming Outdoor Council
201 Main, PO Box 1449, Lander, WY 82520, (307) 332-7031

Wyoming Recreation Commission
122 West 25th St, Cheyenne, WY 82002

Wyoming Trout Ranch
4727 Powell Highway, Cody,
WY 82414

Wyoming Wildlife Federation
PO Box 106, Cheyenne, WY
82003

Wyatt Group, The, Pennsylvania
Environmental Systems Div,
Box 4423, Lancaster, PA
17604, fax (717) 396-9263

Wyman Center
600 Kiwanis Dr, Eureka, MO
63025, (314) 938-5245

Yakima Indian Nation
Box 151, Toppenish, WA
98948, (509) 865-5121

Yanko Environmental Services
3303 Paine Ave, Sheboygan,
WI 53081, (414) 459-2500

YMCA Camp Chingachgook
13 State St, Schenectady, NY
12305

YMCA Camp Cosby
Box 81C, Rte 2, Alpine, AL
35014

YMCA Camp Erdman
PO Box 657, Waialua, HI
96791

YMCA Camp Grady Spruce
Star Route, Box 185, Gradford,
TX 76449, (817) 779-3544

YMCA Camp Jordan
127 Hammond St, Bangor, ME
04401

YMCA Camp Ocoee
301 West 6th St, Chattanooga,
TN 37402, (615) 265-0455

YMCA Camp Ohiyesa
7300 Hickory Ridge Rd, Holly,
MI 48442

YMCA Camp Orkila
Box 1149, Eastsound, WA
98245, (206) 376-2678

YMCA Camp Seymour
9725 Cramer Rd KPN, Gig
Harbor, WA 98329, (201) 884-
3392

*YMCA Camp St. Croix
Environmental Center, Wisconsin*
532 County Rd F, Hudson, WI
54016, (715) 386-2662

YMCA Camp U-Nah-Li-Ya
13654 South Shore Dr, Suring,
WI 54174

YMCA Camp Winona
825 Derbyshire Rd, Daytona
Beach, FL 32055

YMCA Camping Services
350 North 1st Ave, Phoenix,
AZ 85003

YMCA Family Outdoor Center
4540 River Rd, Perry, OH
44081, (216) 259-2724

YMCA Storer Camps
7260 S Stony Lake Rd, Jackson,
MI 49201, (517) 536-8607

YMCA Wilson Outdoor Center, OH
2330 County Rd 11,
Bellefontaine, OH 43311,
(513) 593-6194

YO Ranch Adventure Camp
Mountain Home, TX 78058,
(512) 640-3220

York Services Corp
1 Research Dr, Stamford, CT
06906

Yosemite Institute
PO Box 487, Yosemite, CA
95389

*Youth for Environmental Sanity,
YES!*
706 Frederick St, Santa Cruz,
CA 95062

Youth Service Task Force
Chatham County Human
Resource Dept, 133
Montgomery St, Rm 605,
Savannah, GA 31401

*Zoetic Research Sea Quest
Expeditions*
Box 2424, Friday Harbor, WA
98250, (206) 378-5767

Zero Population Growth
1400 16th St NW, Ste 320,
Washington, DC 20036

Zoological Society of Philadelphia
3400 W Girard Ave,
Philadelphia, PA 19104

Zoovival Trust
Personnel Dept, PO Box
15007, Clearwater, FL 34629

EARTH WORK
RECOMMENDED BOOK LIST

The Conservation Directory, published annually by the National Wildlife Federation. The best overall directory of environmental and conservation organizations. Includes descriptions, addresses, telephone numbers, and names of officials of nonprofits and government agencies on the federal, state, and local levels.

Ecopreneuring: The Complete Guide to Small-Business Opportunities from the Environmental Revolution, by Steven J. Bennett (John Wiley and Sons, 1992). A crash course in starting a green business, includes a resource directory.

The Environmental Career Guide, by Nicholas Basta (John Wiley Press, 1992). Descriptions of many specific environmental professions, tips on entering the job market, and list of resources, including nonprofits, government agencies, and graduate schools of environmental engineering.

Environmental Careers: A Practical Guide to Opportunities in the '90s, by David J. Warner (CRC Press/Lewis Publishers, 1992). Descriptions of specific careers in environmental protection, environmental education, natural resource management, and nondegree technical careers, with an analysis of the best job prospects.

The Environmental Economic Revolution, by Michael Silverstein (St. Martin's Press, 1993). A description from the business view of "the great green restructuring" and green trades.

Environmental Jobs for Scientists and Engineers, by Nicholas Basta (John Wiley and Sons, 1992). Basta is senior editor of *Chemical Engineering* magazine.

A Fierce Green Fire, by Philip Shabecoff (Hill and Wang, 1993). Insightful analysis of the environment movement today by former *New York Times* environmental reporter, essential background for people considering careers in the environmental movement.

Green at Work, by Susan Cohn (Island Press, 1992). Business careers that work for the environment, including opportunities in management and marketing. Includes corporate directory.

The Green Supermarket Shopping Guide, by John F. Wasik (Warner Books, 1993). Contains a rating of 220 corporations and two thousand products by the New Consumer Institute. Directory of company names, addresses, and telephone numbers, and explanations of which companies have taken environmental pledges such as the CERES principles. Could be useful to those interested in green marketing.

Inside the Environmental Movement: Meeting the Leadership Challenge, by Donald Snow, the Conservation Fund (Island Press, 1991). A must-read for those who want to work in the nonprofit conservation community or have a concern for higher education.

The New Complete Guide to Environmental Careers, by the Environmental Careers Organization (formerly CEIP Fund; Island Press, 1993). Complete overview of careers in natural resources, environmental protection, and planning and communication. Kevin Doyle of ECO provides advice regularly in *Earth Work.*

1993 Earth Journal: Environmental Almanac and Resource Directory, annual from the editors of *Buzzworm* (now *Earth Journal)* magazine (Boulder, CO: Buzzworm Books). Includes the year in review plus sections on environmental developments in arts and entertainment, the home, health, gardening, green business, and travel and an environmental education directory.

The 1993 Information Please Environmental Almanac, compiled annually by the World Resources Institute (Houghton Mifflin). An invaluable reference on the state of the planet and an overview of environmental issues, developments, and trends on the local, state, and global levels. Includes state profiles.

Outdoor Careers, by Ellen Shenk (Stackpole Books, 1992). Descriptions and resource lists for jobs ranging from agriculture and food production to biological sciences, parks and recreation, forestry, and marine sciences.

Voices from the Environmental Movement, by the Conservation Fund, edited by Donald Snow (Island Press, 1992). Noted environmentalists speak out in chapters on the environmental movement as a political force, the role of women, international leadership, conservation in academia, and the role of minorities.

CONTRIBUTING EDITORS
AND WRITERS

The Student Conservation Association gratefully acknowledges the work of the coeditors and writers who contributed to this book. Many of the pages in this book were adapted from articles that first appeared in *Earth Work* magazine, published by SCA.

Joan Moody was founding editor of *Earth Work* magazine and coined its name, as well as preparing its contents. She also conceptualized, wrote, and edited much of this book based on her two decades of work in conservation. She is the author of chapter 1 on the job hunt, of chapter 6 on nonprofit organizations, of four sections on federal agencies in chapter 5, of the analysis of "green-hot" jobs in chapter 8, and of miscellaneous chapter introductions, earth worker profiles, and the recommended reading list. She is coauthor of chapter 3 on graduate schools and of the section on international careers. Moody currently is media director for Defenders of Wildlife. She previously served as press secretary and legislative assistant to the late conservationist Representative Phillip Burton; senior editor of *National Parks* magazine; and a public affairs specialist for a number of nonprofit conservation organizations.

Dr. Richard Wizansky served as project director and coeditor for this book. He is development director of the Student Conservation Association, which publishes *Earth Work* magazine. Wizansky holds an Ed.D. in Educational Policy and Research from the University of Massachusetts at Amherst. He wrote the section on the Student Conservation Association and the introduction to chapter 9, and he coauthored the section on the U.S. Forest Service. Dr. Wizansky previously has authored numerous articles on lifestyles and literature.

John Esson is author of the piece on environmental recruiting, the federal salary survey, and the sections on how to secure a federal environmental career, how to get a summer job, and how to find work at the Environmental Protection Agency. Esson is the Director of the Environmental Career Center in Hampton, Virginia, and a frequent contributor to *Earth Work* magazine. Mr. Esson is also an environmental consultant with fifteen years of experience.

Kevin Doyle wrote chapter 7 on environmental management jobs, which first appeared in *Earth Work* Magazine. Doyle is national director of program development for the Environmental Careers Organization, a regular columnist for *Earth Work* magazine, and one of the primary authors of *The New Complete Guide to Environmental Careers*.

Susan Cohn wrote the section on networking in chapter 2. She also wrote the section on careers in green marketing. Cohn is the author of *Green at Work* and serves as a career counselor at New York University's Leonard N. Stern School of Business, where she specializes in green marketing and environmental opportunities related to business corporations.

T. Destry Jarvis is author of the profile of the Environmental Careers Organization. As executive vice president of SCA, he expanded SCA's career services and was publisher of *Earth Work* magazine. His vision also led to the founding of the Conservation Career Development Program. He currently is special assistant for policy to Rodger Kennedy, the director of the National Park Service.

Marta Cruz Kelly wrote the section on diversity in conservation. She is director of SCA's Conservation Career Development Program, a former National Park Service ranger, and a trainer in cultural diversity and interpersonal relations.

Douglas Fulmer is coauthor of chapter 3 and wrote the sections on the Peace Corps and the U.S. Fish and Wildlife Service. He is a Baltimore-based freelance writer and political consultant whose environmental work has appeared in *Earth Work* and many other publications.

Jonathan B. Jarvis, author of "So You Want to be a Park Ranger?" is superintendent of Craters of the Moon National Monument in Idaho.

Mary McKelvey coauthored the section on international environmental careers. Fluent in Russian, she served as a Student Conservation Association supervisor in an exchange program in U.S. and Russian parks.

Nicholas Basta wrote the section on environmental engineers, which appeared as an article in *Earth Work* magazine. He is author of *Environmental Jobs for Scientists and Engineers* and senior editor of *Chemical Engineering* magazine.

Karen Tiemens wrote the sections on biodiversity and natural resource damage assessment in chapter 8. She is a freelance medical and scientific writer based in Falls Church, Virginia.

William Thomas Barrett, J.D., of the University of Vermont School of Law, wrote the section on careers in environmental law.

Nat Bulkley wrote the section on green teachers, adapted from *Earth Work* magazine.